IN CASE OF ACCIDENT[1]

*In case of accident notify the laboratory instructor **immediately**.*

Fire

Burning Clothing. If a person's clothing is on fire, douse the individual with water or wrap the person in a coat, blanket or whatever is available to extinguish the fire, or roll the person on the floor, if necessary. Quickly remove any clothing contaminated with chemicals. Use caution when removing pullover shirts or sweaters to prevent contamination of the eyes. Douse with water to remove heat and place clean, wet, ice-packed cloths on burned areas. Wrap the injured person to avoid shock and exposure. Get medical attention promptly.

Burning Reagents. Extinguish all nearby burners and remove combustible material and solvents. Small fires in flasks and beakers can be extinguished by covering the container with an asbestos-wire gauze square, a big beaker, or a watch glass. Use a dry chemical or carbon dioxide fire extinguisher directed at the base of the flames. **Do not use water.**

Burns, either Thermal or Chemical. Flush the burned area with cold water for at least 15 min. Resume if pain returns. Wash off chemicals with a mild detergent and water. Current practice recommends that no neutralizing chemicals, unguents, creams, lotions, or salves be applied. If chemicals are spilled on a person over a large area quickly remove the contaminated clothing while under the safety shower. Seconds count and time should not be wasted because of modesty. Get prompt medical attention.

Chemicals in the Eye

Flush the eye with copious amounts of water for 15 min. using an eye-wash fountain or bottle, or by placing the injured person face up on the floor and pouring water in the open eye. Hold the eye open to wash behind the eyelids. After 15 min. of washing obtain prompt medical attention, regardless of the severity of the injury.

Cuts

Minor Cuts. This type of cut is most common in the organic laboratory and usually arises from broken glass. Wash the cut, remove any pieces of glass, and apply pressure to stop the bleeding. Get medical attention.

Major Cuts. If blood is spurting place a pad directly on the wound, apply firm pressure, wrap the injured to avoid shock, and get **immediate** medical attention. Never use a tourniquet.

Poisons

Call 800 information (1-800-555-1212) for the telephone number of the nearest Poison Control Center, which is usually an 800 number also.

[1]Adapted from *Safety in Academic Chemistry Laboratories*, prepared by the American Chemical Society Committee on Chemical Safety, March 1974 and 1985.

Organic Experiments

Sixth Edition

Walter W. Linstromberg
University of Nebraska at Omaha

Henry E. Baumgarten
University of Nebraska–Lincoln

D. C. HEATH AND COMPANY   Lexington, Massachusetts Toronto

Preface

The sixth edition of this manual reflects much the same intended audience, organization, and emphasis as in previous editions. It is intended primarily for use in introductory organic chemistry courses having fifteen to thirty laboratory periods. Since there is a considerable variation in the equipment and facilities available to users of the manual, we have attempted to provide a selection of experiments that can be carried out in almost any contemporary elementary organic chemistry laboratory. Most of the early experiments in the manual are "technique" experiments which introduce students to the fundamental operations and techniques that form the basis for most laboratory work in organic chemistry. These introductory experiments are followed by test tube, simple preparative, and characterization experiments that are arranged more or less in the order usually followed in introducing functional groups. Following these is a series of preparative experiments chosen from, but not limited to, the principal synthetic reactions that form the basis of typical introductory courses. Several of the experiments may also be run in combination with other experiments to form short reaction sequences illustrative of more advanced syntheses.

In this revision our efforts have been directed largely toward improving the core experiments in the manual, that body of experiments performed by almost all users. To accomplish this, many of the early experiments have been revised substantially. However, all experiments have been reexamined, particularly with regard to safe laboratory practices. The authors believe that the experiments in this manual are no more hazardous than those usually carried out in beginning student laboratories; however, in accord with almost universal present practice, we have considerably increased the number and scope of warning notices and caution alerts that inform students and instructors of potential hazards, however remote these dangers may be. In accordance with our past practice, we have avoided the use of *carbon tetrachloride, chloroform,* and *benzene* in the experiments. As this was being written, the status of methylene chloride as a possible cancer suspect agent was under debate. Users of this manual will want to familiarize themselves with the most current information about the status of this solvent, which is used in several experiments, and select experiments based on their best judgment.

Previous users of the manual will find four new or greatly expanded introductory sections: (1) Laboratory Safety; (2) Calculation of Yields; (3) Laboratory Reports; and (4) Methods of Heating. The revised section on laboratory safety is provided because of our common and continuing concern for promoting and encouraging safe practices in the laboratory. To this end, we have categorized all reagents and chemicals that must be handled with caution. The new section on calculation of yields and the expanded section on laboratory reports and notebooks were added in direct response to the requests of several users of long standing. The new section that describes the various methods of heating was added in answer to questions from a number of users who prefer, or are limited to, specific methods of heating. Although there is probably no single method of heating that is uniformly good for all experiments, we believe that, with due care, most of the common methods can be used for the experiments in this manual.

The sixth edition includes five new experiments intended to offer a more complete selection. Experiment 6 illustrates the use of sublimation as a method of purification and involves the isolation of caffeine from tea, followed by sublimation of the product. Experiment 10 requires students to construct molecular models of a number of compounds, including isomers of various sorts. Experiment 18 is a collection of three procedures with the central theme of stereochemistry. In the first part of this experiment, the isomerization of maleic acid to fumaric acid is carried out. In the second part, L-3-phenyllactic acid is prepared by the stereoselective diazotization of L-phenylalanine under van Slyke conditions. In the third part, ethyl acetoacetate is reduced enzymatically and stereospecifically by yeast to ethyl (S)-3-hydroxybutanoate.

As was previously mentioned, several experiments have undergone substantial revision. For some experiments the revision required the addition of new parts or the replacement of old procedures by new ones. Experiments so changed include numbers 6, 8, 13, 14, and 20.

Special care continues to be given to the economy of materials used in the experiments. Wherever possible, reagents, chemicals, and unknowns of relatively low cost have been selected, with the quantities used kept as small as is practical with the commonly available apparatus. In a few experiments, a more expensive reagent has been used, because it is demonstrably either much better or much safer than possible alternatives. In sequential experiments, examples have been chosen such that all intermediates are either prepared by the student or commercially available at a relatively low cost.

Estimated time schedules are included for all experiments. These times should be considered as only rough guides, since efficiency in the laboratory varies with the individual student and the equipment available. The schedules are estimates based on experience with both majors and nonmajors under reasonably close supervision. In the process we have observed that certain items of equipment or facilities appear to function as bottlenecks in the organic teaching laboratory: balances, melting point apparatuses, polarimeters, gas chromatographs, spectrometers in general, and fume hoods. With large sections and limited special equipment, more time may be required. When preparations require longer periods than allowed for in a typical laboratory session, a safe point at which the experiment may be interrupted is cited in the procedure.

New to this edition is an Instructor's Guide. This guide is intended to aid instructors, laboratory assistants, and storeroom personnel, all of whom are concerned with the selection of, preparation for, and supervision of experiments from the manual. It includes notes, comments, and suggestions for the individual experiments; answers to all of the exercises; and lists of equipment, chemicals, and supplies for the experiments. In the process of preparing the guide we have added new exercises and pruned the old ones. The exercises vary in difficulty and are intended to explore many facets of chemistry, including the technique, mechanism, stoichiometry, and general knowledge required by the procedure under study. Some problems require the use of ir and nmr spectra for their solution; these are marked with an asterisk.

The authors appreciate the continued wide acceptance of the previous editions and are grateful for the many helpful comments, suggestions, and questions that have been sent to us by users. Many of the changes made in the sixth edition originated from suggestions or questions from users—to these people the authors are greatly indebted. Two of our colleagues have been especially helpful to the authors in the course of this revision; James K. Wood, a very dedicated teacher at the University of Nebraska at Omaha, and Gholam Mirafazl, an outstanding teaching assistant at the University of Nebraska–Lincoln.

The authors are also grateful to the following for their help: Alexander G. Bednekoff, Pittsburgh State University; W. H. Bogart Jr., Danville State College; Warren Bosch, Elgin Community College; Robert Boxer, Georgia Southern College; James Brown, Middle Tennessee State University; K. M. Elsen, Mount Mary College; James D. Fisk, Samford University; Patrick M. Garvey, Des Moines Area Community College; Estelle Gearon, Montgomery College–Takoma Park; George J. Gradel, Philadelphia College of Textiles and Science; Richard Hoffmann, Illinois Central University; Elmer E. Jones, Northeastern University; Martin B. Jones, University of South Dakota; John Kenkel, Southeast Community College; Anne Loeb, College of Lake County; Daniel O'Brien, Texas A & M University; Kenneth Pohlmann, St. Mary's University of San Antonio; Jim Schammerhorn, Murray State College; Charles E. Sundin, University of Wisconsin–Plattville; Lynn G. Wiedermann, Black Hawk State College; and Charles D. Yates, State Technical Institute at Memphis.

As is our custom, we close with an invitation to all users to consider this collection of experiments as their manual, to feel free to improvise and improve upon it, and to send their findings, comments, and questions to the authors, because all successful organic chemistry laboratory manuals represent the contributions of many teachers, past and present.

WALTER W. LINSTROMBERG
HENRY E. BAUMGARTEN

Contents

Experiments

Appendices

General Instructions

The purpose of laboratory work in organic chemistry is twofold: (1) to illustrate the reactions and general principles discussed in lectures and your textbook and (2) to acquaint you with those laboratory practices and techniques that make up what has been called the **art of organic chemistry.** You soon will realize that reactions in organic chemistry are far different from those previously encountered in your general, or inorganic, chemistry course. Organic reactions frequently require long periods of time (even at elevated temperatures) for completion, and complicating side reactions that lead to a mixture of products are quite common. Considerable manipulation and treatment are required for the isolation of most pure products, and final yields may be much less than those theoretically possible. In general, the acceptability of an organic reaction is governed not only by the quantity of product obtained from the reaction but also by the quality (purity) of the product. A high yield of an impure substance is no better and, in some cases, is far worse than a low yield of a pure product.

For all the seemingly unkempt appearance of most organic chemistry laboratories, no branch of chemistry requires a higher development of laboratory techniques or a greater refinement of laboratory practices than organic chemistry. The organic chemist has developed a number of laboratory operations and many simple pieces of laboratory equipment that will be unfamiliar to you but that have been found to be indispensable in the practice of this science. Learning proper procedures and techniques in the organic laboratory will aid you in other science courses that follow, especially those in the medical, pharmaceutical, and biological sciences.

Follow directions carefully, ask questions whenever the procedure is not clear to you, and exercise care in the purification of any products made. These are the simple directions that will lead to a proper achievement of the purpose of the laboratory. You are expected to work alone unless otherwise instructed. Address all questions concerning the experiments to your laboratory instructor, *not to your neighbor*. Results, when obtained, or observations, when made, should be recorded immediately, either on the report form provided or in your laboratory notebook depending upon the directions your instructor gives. Complete all experiments with the exception of multi-step preparations within the prescribed laboratory period. Clean all apparatus at the *end* of each experiment rather than before. It is always more easily cleaned after recent use; moreover, it is better to begin an experiment with clean and *dry* apparatus.

Laboratory Safety

How safe is your *home?* It is instructive to make a list of all of the hazardous materials found in an ordinary household: medicinal products; insecticides; adhesives; ammonia and household bleach; detergents and spot removers; painting supplies; natural gas or propane; a host of electrical appliances and the associated wiring; oils and gasoline for the mower or boat; hobbyist supplies for photography, art, and crafts; glass containers, cookware, and dishes; windows, knives, razors, scissors—the list goes on and on. When you do this it will come as no surprise to find that there are some hazards in the chemistry laboratory also. Those of us who are in the chemical and biological professions must learn to work safely with the hazards of our workplace, as we all must learn to live with the hazards of our homes, transportation systems, and places of recreation. As you learn about the laboratory techniques of organic chemistry, you will find that your instructor will place great emphasis not only on using each of these techniques properly but also on using them safely. In the organic chemistry laboratory, as in any chemical, biological, or medical laboratory, you will have to learn to work with fragile glass apparatus, with flammable liquids and solids, and with corrosive or toxic chemicals. Although the experiments in this manual have been chosen to minimize the use of hazardous materials and equipment, it is impossible to avoid these entirely. Indeed, most experiments in the organic chemistry laboratory involve some chemical that could be regarded as hazardous: acids, bases, solvents, etc. Therefore, it is up to you to learn the essential rules for laboratory safety and to carry out each of your experiments in such a way as not to endanger yourself or your fellow students. Read each experiment carefully and completely before coming to the laboratory, noting especially any safety precautions given in this manual. If you have any questions about the safety of the procedures involved, ask your instructor about these before beginning the experiment. If you follow safe procedures, you will find that the organic laboratory is no more—probably less—dangerous than your kitchen, workshop, or garage. In the following paragraphs, we will consider some guidelines for laboratory safety. Some of these are summarized on the inside front cover of this manual. However, because of the great variation among chemistry laboratories, your instructor may have alternative or additional safety rules for you to follow.

Learn the safety rules for your laboratory. Your instructor will have procedures for the handling of accidents and for the evacuation of the laboratory in the event of fire or other hazardous events. Learn what these procedures are before you or your classmates need them.

Wear safety goggles or approved safety glasses in the laboratory at all times, whether you are working at the bench, writing in your notebook, or washing glassware. You could be the innocent victim of someone else's accident. Your eyes are especially vulnerable to damage by chemicals or flying glass particles, and some eye damage can be irreversible. **Do not wear contact lenses in the laboratory.** You may not be able to get a lens out fast enough to prevent eye damage if a hazardous chemical gets beneath the lens. If you get any chemical in your eyes, flush your eyes thoroughly with water (at an eye-wash fountain, if available) for at least fifteen minutes and seek immediate medical attention.

Check all apparatus before use, after assembly, and regularly during each experiment. Begin doing this at the time you first check into the laboratory. Examine all of your glassware for cracks, "stars," or other signs of weakness and have the storeroom replace faulty glassware. Learn to do this routinely each time you assemble glass apparatus for an experiment. Do not use electrical equipment with frayed cords or suspicious connections. Make certain that all assemblies of apparatus are mechanically sound and stable, not flimsy or poorly supported. Understand the function of any piece of equipment or glassware *before* you begin to use it.

Follow safe disposal practices for chemical wastes. Your instructor will provide you with instructions for the disposing of both liquid and solid wastes from your experiments in such a way as to protect the environment and those who clean the laboratories. Most aqueous solutions can be flushed down the drain with much water, but always check with your instructor before doing so. In general, it is not wise to mix different types of organic liquid wastes or to put organic solids in with the ordinary trash. Proper disposal of organic and inorganic residues is a part of every chemical experiment.

Handle all flammable materials with extreme care. One of the principal hazards in the organic chemistry laboratory is fire. Many organic chemicals are highly flammable and most organic liquids and vapors will burn when exposed to high heat or an open flame. Although it is considered good practice, insofar as possible, to avoid the use of burners and open flames in the organic chemistry laboratory, in many student laboratories Bunsen burners have been and are being used *safely* as the principal source of heat. However, whenever possible use heating mantles, oil baths, steam baths, or hot plates with due regard for the safety factors associated with these devices. If it is necessary to heat a flammable liquid with a burner, the liquid must be contained in a closed vessel protected with a reflux condenser. Do not heat any glass vessel with a burner over a plain wire gauze. The gauze can cause hot spots to develop that may lead to cracking of the vessel. Use instead a ceramic fiber-centered gauze—*not* asbestos-centered!—between the vessel and the flame. Never heat any liquid in an open vessel unless you are specifically told to do so (as in a melting point determination). Before transferring, pouring, measuring, or otherwise handling a flammable liquid, be sure that there are no lighted burners nearby. Before lighting a burner make certain that none of your neighbors is working with a flammable liquid in an open vessel. Do not pour organic solvents down the drain as their vapors could move through the drainage system to other areas where there are flames.

Avoid contact of chemicals with the skin or other tissues. It is now considered good practice to wear disposable plastic gloves in the laboratory to minimize the contact of chemicals with the skin whenever possibly toxic chemicals must be used. These gloves are rather fragile and should be replaced immediately if torn or contaminated with any toxic chemical. For cleaning up anything more than a minor spill, wear sturdy protective gloves. Wear shoes that cover your toes and instep. Be especially careful while transferring both liquids and solids from one container to another, or while weighing out chemicals of any type. Keep the area around balances used for weighing chemicals clean and free of chemical residues. If you get any chemical on your skin, wash the area thoroughly with soap and water. Do not use organic solvents to cleanse your skin of possibly toxic substances. Organic solvents can cause the chemicals to penetrate into and beneath the skin.

Avoid inhaling vapors. Whenever possible, use a good fume hood to carry out all operations involving compounds with toxic vapors and all reactions that give off toxic or irritating vapors. This includes not only organic vapors but vapors from inorganic substances such as bromine, hydrogen chloride, ammonia, and thionyl chloride. If you should inhale any suspicious vapors or gases, notify your instructor at once.

Handle all chemicals as though your safety depends upon proper technique—as well it may. Do not inhale, taste, or smell any chemicals unless you are specifically told to do so *by your instructor*. Most chemicals (even some of those you commonly eat or drink) are mildly toxic. Others may be very poisonous, caustic, corrosive, carcinogenic, explosive, or flammable.

Every effort will be made to warn you about the hazardous properties of any chemicals used in your experiments. However, our knowledge of all of the hazards associated with chemicals is still imperfect. Therefore, it is considered good practice to treat *all* chemicals as if they were potentially hazardous.

A list of chemicals and reagents that are real or suspected hazards in the organic chemistry laboratory follows. The substances are listed under specific categories where their hazardous properties are described and directions for safe handling are given. Some chemicals fall into more than one hazard category. Many experiments have a preface in which the possibly hazardous chemicals are listed with references to the following descriptions. Before beginning any experiment look for such a preface, and, if there is one, read the appropriate hazard descriptions.

You will also find numerous caution signs and caution notices appearing in the text itself. The signs resemble the standard Department of Transportation (DOT) hazard labels that are widely used throughout chemistry in the same spirit as the standardized highway markers used to warn motorists of curves, hills, and dangerous intersections. The labels are intended to alert you to possible hazards if good safety practices are not followed. In many laboratories, the appropriate DOT labels are routinely put on all bottles of chemicals in the teaching laboratories. Exercise care and develop a second sense for safety. You may find that the safety skills that you acquire in the organic chemistry laboratory will make your home a safer place as well.

Hazard Categories

1. *Flammable Reagents*

   Ether. Diethyl ether is very flammable. It is also an anesthetic and can have a narcotic effect if inhaled in too great a quantity. Always use ether with adequate ventilation in an area well removed from flames.

   Liquid hydrocarbons. Petroleum ether (a hydrocarbon mixture) and all of the common *alkanes, alkenes, alkynes,* and *aromatic* hydrocarbons are highly flammable and have low flash points. Handle these materials with the same care as you would use with ether.

   Organic solvents. Organic solvents other than ether and the alkanes include *acetone,* the *alcohols,* and a number of *esters.* Handle these materials with the same care as you would use with ether.

   Tetrahydrofuran. Tetrahydrofuran is a cyclic ether and very flammable. Use the same precautions with this reagent as you would use with diethyl ether.

2. *Corrosive Reagents*

   Hydrochloric acid. Both the liquid and the vapors of hydrochloric acid are very corrosive. Avoid contact. Avoid inhalation of vapors. Spills of this acid on equipment or any part of the body should be washed off immediately with large amounts of water.

Sulfuric acid. Sulfuric acid is a very corrosive liquid that causes serious acid burns. The eyes especially are vulnerable to sulfuric acid burns. Avoid contact. Wear your safety glasses! Spills should be washed immediately with large amounts of water.

Sodium hydroxide. Sodium hydroxide and its solutions are corrosive. Wash off any spills immediately with water until the affected areas no longer feel slippery to the touch.

Ammonium hydroxide. Ammonium hydroxide is both corrosive *and* toxic. Avoid contact. Avoid inhalation of vapors. Wash off any spills with large quantities of water.

Nitrating mixtures. Mixtures of nitric and sulfuric acid are not only very corrosive, but are also strong oxidizing and dehydrating agents. These mixtures are best used in a good exhaust hood. Avoid contact. Avoid inhalation of fumes. Wear protective gloves. Wash off any spills immediately with large amounts of water.

Soda-lime. Soda-lime is a mixture of sodium hydroxide and calcium oxide. Not only is soda-lime corrosive, but it is also a powerful dessicant. Avoid contact. Treat it as you would sodium hydroxide.

Phosphoric acid. Phosphoric acid is a very corrosive liquid. Employ the same precautions that you would use with sulfuric acid.

Maleic anhydride. Maleic anhydride reacts with water to form maleic acid, a somewhat stronger and more corrosive chemical than most organic acids. Avoid contact. Wash off spills with copious amounts of soap and water.

Hydrobromic acid. Both the liquid and vapors of hydrobromic acid are very corrosive. Avoid contact. Use the same care in working with hydrobromic acid as you would in working with hydrochloric acid.

3. *Toxic Reagents*

Organic Amines. Exercise care in working with organic amines, especially aniline, aniline derivatives, and other aromatic amines. Many organic amines are quite toxic. Some are cancer suspect and nearly all are malodorous. Avoid contact with these materials with any part of the body since some of them can be absorbed through the skin. Avoid inhalation of their vapors. Wash off any particles or solutions of these materials that come in contact with the skin with generous amounts of soap and water.

Phenols. Phenols, especially the low molecular weight, water-soluble, volatile phenols are toxic and/or corrosive. Avoid contact. Wash off any particles or solutions of phenol with generous quantities of soap and water.

Bromine. Bromine is one of the most pernicious substances used in the organic chemistry laboratory. Not only is bromine highly toxic, but its vapors are extremely irritating to the eyes, nose, and mucous membranes. Bromine and its solutions should be dispensed and transferred in a good exhaust hood. Wear plastic gloves while working with bromine or its solutions. If you should come into contact with bromine, wash it off immediately with soap and water, then apply a gauze dressing soaked in 10% sodium thiosulfate solution for several hours. Unless the area is small, it may be wise to see a physician.

Methanol. Methanol (methyl alcohol) is very toxic. Avoid contact with the skin. Avoid breathing the vapors. Wash off spills with generous quantities of soap and water.

Mercuric chloride. Mercuric chloride and other divalent mercury compounds are very toxic. Avoid contact. Wash off spills with generous quantities of soap and water.

4. *Carcinogenic Reagents*

Dyestuffs. Rhodamine, along with a number of other organic dyes, is currently listed as cancer suspect. Moreover, most dyes can also color your skin and clothing. Avoid contact. Wear plastic gloves and a laboratory apron. Wash off spills with generous amounts of soap and water.

Potassium dichromate. Potassium dichromate ($K_2Cr_2O_7$) and other hexavalent chromium compounds are cancer suspect agents. Avoid contact. Wash off any particles or solutions of chromium oxidants that come in contact with any part of the body with generous quantities of soap and water. Do not discard hexavalent chromium compounds in the sink. Reduce hexavalent chromium compounds to the trivalent state before discarding. Follow the instructions provided by your instructor.

Chlorinated hydrocarbons. Certain of the alkyl chlorides and bromides are cancer suspect. Handle with care and avoid contact.

Hydrazine. Hydrazine and hydrazine derivatives are thought to be carcinogenic. Handle with care and avoid contact. Wash off any particles or solutions of hydrazine derivatives with liberal use of soap and water. Report any spills to your instructor.

Formaldehyde. Formaldehyde is a toxic and cancer suspect gas. It is marketed as a 35–40% aqueous solution known as "formalin." Try to avoid contact with formalin solution, and avoid inhalation of formaldehyde vapors. Formaldehyde should be dispensed and used in a good fume hood.

Tannic acid. Tannic acid is currently listed as cancer suspect. Avoid contact. Wash off spills with copious amounts of soap and water. Report any spills to your laboratory instructor.

5. *Lachrymatory Agents*

Iodine. Iodine is a lachrymator and is also corrosive. Avoid contact. Avoid inhalation of vapor.

Phenyl isothiocyanate. Phenyl isothiocyanate is a lachrymator and a malodorous reagent. Avoid contact of the vapors with the skin, eyes, nose, and mucous membranes. This reagent should be dispensed and used in a good exhaust hood.

6. *Explosive Reagents* (*See also Note 1, p. 57.*)

Sodium. Sodium metal reacts violently with water. Therefore, all apparatus used with sodium must be clean and dry. Never dispose of sodium in the sink! Unreacted sodium is best destroyed by allowing it to react with ethanol until no more hydrogen evolution is evident. The ethanolic solution of sodium alcoholate may be discarded in the sink with large quantities of water.

Picric acid. Picric acid (2,4,6-trinitrophenol) is explosive as well as toxic. It is marketed containing about 35% water, and should be kept in a slightly moistened condition.

Silver acetylides. Silver acetylides are shock-sensitive when dry.

7. *Malodorous Reagents*

Aldehydes. The lower molecular weight aldehydes, especially propionaldehyde and butyraldehyde, are very unpleasant and difficult to eliminate if absorbed by the skin or clothing. Avoid spilling these materials. Avoid inhaling the vapors. Wash off spills with generous quantities of soap and water.

Pyridine. Avoid contact. Avoid inhalation of vapors. Wash off spills with generous use of soap and water.

Phenylacetic acid. Phenylacetic acid is an extremely obnoxious reagent with an odor that infects the skin, hair, clothing, books, and desk top. Avoid contact. Avoid spilling. Wear plastic gloves. Wash spills with a sodium bicarbonate solution, soap, and water.

8. *Irritants*

Acyl halides and anhydrides. Liquid acyl halides and anhydrides such as acetyl chloride, benzoyl chloride, and acetic anhydride are very irritating to the skin and mucous membranes. Avoid contact. Avoid breathing the vapors of these substances. Wash off any of these materials or their solutions that come into contact with your body with copious amounts of soap and water.

Organic acids. Some organic acids are strong irritants and corrosive as well. Avoid contact. Wash off spills with generous amounts of soap and water.

Pyridine. Pyridine is classified as a strong irritant. It is also toxic and malodorous. Avoid contact. Avoid breathing its vapors. Wash off any spills with generous amounts of soap and water.

2-Naphthol. 2-Naphthol is toxic and an irritant. Avoid contact. Wash off spills with copious quantities of soap and water.

Tetrahydrofuran. Tetrahydrofuran is an irritant. Avoid contact. Wash off spills with copious quantities of soap and water.

Thionyl chloride. Thionyl chloride is a particularly strong irritant that fumes in moist air to liberate sulfur dioxide and hydrogen chloride—both strong irritants. Avoid contact. Avoid inhalation of fumes. Wash off spills with copious amounts of soap and water.

Calculation of Yields in Synthesis

In some of the experiments in this manual you will prepare or synthesize one or more organic products from one or more organic reactants. In all laboratory procedures that lead to the preparation of an organic compound, the amount of product of acceptable purity actually obtained is called the **yield.** The yield is dependent upon a number of factors, some of which are (1) the relative amount of each reactant, (2) the length of time we allow for the reaction, (3) the temperature at which the reaction takes place, (4) the possibility of side reactions giving unwanted byproducts, and (5) the mechanical losses inherent in the purification process. Thus, very few organic syntheses result in the complete conversion of all of the starting material to the desired product. Even if this were to happen, some of the product would almost inevitably be lost during isolation and purification. A simple way of stating this observation is to say that the **percentage yield** of most organic reactions is less than 100%, where the percentage yield is the ratio of the amount of product actually obtained to that theoretically possible × 100. To calculate the percentage yield, we must first calculate the **theoretical yield,** the maximum yield that could be obtained with the quantities of reactants used in the experiment.

Before discussing the calculation of theoretical yields, we will review how chemists *count* molecules. Dogs, cats, oranges, apples, boys, girls—these are easily counted directly. Molecules, however, are extremely small, and we have no easy way to count molecules of a substance directly. Therefore, we count them indirectly by weighing the substance. Thus, you have learned that one gram-molecular weight, one molecular weight in grams, of a substance is equal to 6.02×10^{23} (Avogadro's number of) molecules of the substance. We call this number of molecules a **mole** (or gram mole). Thus, in carrying out the experiments in this manual, when you weigh (or measure by volume) an organic chemical or reagent, you are counting the molecules of that substance or reagent. You will want to do so carefully, and we urge you to try to carry out all weighing (counting) with reasonable accuracy: to about ± 0.1 g for most preparative experiments.

To calculate the theoretical yield of a reaction we begin by writing a *balanced* equation for the reaction. Consider, first, a reaction whose stoichiometry is shown by the simple balanced equation:

$$A + B \longrightarrow C + D$$

What does this equation tell us? Using the commonly accepted chemical conventions that you studied in general chemistry, it tells us several things, depending on how we intend to use it. When we are studying organic reactions, it tells us that one molecule of A reacts with one molecule of B to yield one molecule of C and one molecule of D. This is important information, but in the laboratory we do not work with individual molecules of A, B, C, and D but rather many molecules of A, B, C, and D. Therefore, in the absence of any other information, the equation tells us that *one mole* of A reacts with *one mole* of B to yield *one mole* of C and *one mole* of D.

Of course, if the equation is true for one mole of each reactant and product, it should still be true for half a mole of each or for any specific fraction of a mole of *each* reactant and product. In other words, the equation tells us that the mole ratios for this reaction are: (moles A) : (moles B) : (moles C) : (moles D) = 1 : 1 : 1 : 1.

Next, let us assume that

$$\text{Molecular weight of A} = 60 \text{ g/mole}$$

$$\text{Molecular weight of B} = 80 \text{ g/mole}$$

$$\text{Molecular weight of C} = 100 \text{ g/mole}$$

$$\text{Molecular weight of D} = 40 \text{ g/mole}$$

This means that our still-balanced equation can also be rewritten as

$$60 \text{ g of A} + 80 \text{ g of B} \longrightarrow 100 \text{ g of C} + 40 \text{ g of D}$$

Of course, if we were to divide all of the numbers of grams by 10 (or any other number), the equation would still be balanced. Therefore, still another way of interpreting our original equation is to state that the weight ratios of reactants and products for the balanced equation are:

$$\frac{\text{weight A}}{60} = \frac{\text{weight B}}{80} = \frac{\text{weight C}}{100} = \frac{\text{weight D}}{40}$$

This means that, if we know the weight of any one of the reactants or products, we can calculate the other three using these ratios. Unfortunately, in the organic chemistry laboratory (for reasons that you will learn about in your lectures) we may find it necessary to use more than 80 g of B to react with 60 g of A. We have tried to determine how much of each reactant must be used to get a good yield of product in a reasonable length of time. Thus, you will usually be told how many grams (or milliliters) of each reactant to use, and, from this information, you will be expected to calculate the theoretical yield of the product. You cannot assume that the amounts specified in the experiment are already balanced by weight (or by moles), but you must determine whether there is an excess of any reactant relative to the others. To do this, you may use a simple set of calculations based on the principles that we have just reviewed.

For each reactant we calculate the number of moles that will actually be used in the synthesis using the following familiar equations:

$$\text{Moles of A} = \frac{\text{Weight of A}}{\text{Molecular weight of A}}$$

$$\text{Moles of B} = \frac{\text{Weight of B}}{\text{Molecular weight of B}}$$

If moles of A equals moles of B, then the reactants are balanced, and we can use either moles of A or moles of B to calculate the theoretical yields of products C and D. If moles of A is larger than moles of B, then we have an excess of A present (more A than is needed to react with all of the B present). If moles of B is larger than moles of A, then we have an excess of B. In either case, the theoretical yield of product (C or D) is limited by *the number of moles of the reactant present in smaller amount,* the smaller of the two numbers we obtained in the calculation above. We call this reactant the **limiting reactant.** Next we calculate the theoretical yield of products using the abbreviation MW for molecular weight and the equations:

Theoretical yield of C = (Moles of limiting reactant) × (MW of C)

Theoretical yield of D = (Moles of limiting reactant) × (MW of D)

Finally, after we run the reaction and determine the actual weight of the product, we calculate the percentage yield of the product from the equation:

$$\% \text{ Yield} = \frac{\text{Weight of product actually obtained}}{\text{Theoretical yield of product}} \times 100$$

Before proceeding to some more complex calculations we will examine two examples of simple calculations, one involving only one reactant and a second involving two reactants. Consider first the reaction of 20.0 g of cyclopentanol with a strong acid, such as phosphoric acid, to form cyclopentene and water. We begin by writing the balanced equation for the reaction as shown below. It is best to write both the structural formula and the molecular formula to facilitate calculating the molecular weights, if this is necessary. We obtain the molecular weights of the reactant and expected product from a chemical handbook if we can, or calculate them from the atomic weights if we cannot. Then we write the equation, the quantities of reactants, molecular weights, moles, theoretical quantities of products, etc., in the following format. Note that, although we include the phosphoric acid in the equation, we make no calculations on this acid because it is a catalyst, not a reactant.

		$C_5H_{10}O$ Cyclopentanol			C_5H_8 Cyclopentene		Water
Grams		20.0	10 mL	③ ↑	15.8	④ ↑	4.2
Molecular weight		86.14			68.13		18.015
Moles	①	0.232		②	0.232	②	0.232

The calculated numbers are obtained using the above equations in a stepwise fashion:

Step 1 Moles of cyclopentanol = 20.0/86.14 = 0.232.

Step 2 Since there is only one reactant, it must be the limiting reactant. Therefore, the theoretical moles of product = 0.232. We enter this number in the cyclopentene and water columns.

Step 3   The theoretical yield of cyclopentene = 0.232 × 68.13 = 15.8.

Step 4   Generally we do not calculate yields of relatively unimportant coproducts such as water unless there is a reason to do so. In this example we do so for completeness.

Note the flow of the calculation as indicated by the arrows. As we go down the table, we divide by the molecular weight to obtain the moles. After selecting the limiting (and only) reactant, we go up the table to calculate expected values by multiplying the moles of limiting reactant by the molecular weight of the product. In this manual we will always arrange the columns so that we divide when going down a column and multiply when going up.

In this example both our reactant and product are liquids. We can measure liquids by either weight or volume. However, our calculations are based on weights of reactants and products as well as molecular weights and stoichiometries. Therefore, the following equations for interconverting liquid or solution quantities and mole or weight quantities will prove to be useful.

$$\text{grams of a liquid} = (\text{mL of a liquid}) \times (\text{density of the liquid})$$

$$\text{mL of a liquid} = \frac{\text{grams of a liquid}}{\text{density of the liquid}}$$

$$\text{Moles of a solution reactant} = \frac{(\text{mL of solution}) \times (\text{Molarity of solution})}{1000}$$

Next we consider a slightly more difficult example, a reaction in which there are two reactants, one of which is in excess. Our example is the reaction of 10.0 mL of acetic acid with 20.0 mL of ethyl alcohol to form the ester, ethyl acetate, and water. Note that our starting quantities are given as volumes; therefore, we will have to convert these to grams. First, however, we start by writing the balanced equation; then we prepare a table as before, using the same format and getting as much information as possible from the chemical handbook.

$$\underset{\substack{C_2H_4O_2 \\ \text{Acetic acid}}}{CH_3-\overset{\overset{\displaystyle O}{\|}}{C}-OH} \; + \; \underset{\substack{C_2H_6O \\ \text{Ethyl alcohol}}}{CH_3CH_2OH} \; \xrightarrow{H_2SO_4} \; \underset{\substack{C_4H_8O_2 \\ \text{Ethyl acetate}}}{CH_3-\overset{\overset{\displaystyle O}{\|}}{C}-OCH_2CH_3} \; + \; H_2O$$

Grams	①	9.3 (10.0 mL)	①	15.8 (20.0 mL)	⑤⑥	13.6 (15.1 mL)
Molecular weight		60.05		46.07		88.12
Moles	②	0.154	③	0.343	④	0.154
				→×		

Density	0.9276	0.7893	0.9003

Again the calculated numbers are obtained using the equations given above in a stepwise fashion:

Step 1 Before we begin our calculation we convert all volume quantities to grams: 10.0 mL × 0.9276 = 9.3 g of acetic acid; 20.0 mL × 0.7893 = 15.8 g of ethyl alcohol.

Step 2 Moles of acetic acid = 9.3/60.05 = 0.154.

Step 3 Moles of ethyl alcohol = 15.8/46.07 = 0.343.

Step 4 The smaller number of moles of reactant is seen to be that of acetic acid; therefore, acetic acid is the limiting reactant, and we expect only 0.154 mole of product. We enter 0.154 in the ethyl acetate column.

Step 5 The theoretical yield of ethyl acetate = 0.154 × 88.12 = 13.6 g.

Step 6 Although you will usually report your yields by weight rather than by volume, as an illustration we carry out the conversion. Theoretical volume of product = 13.6/0.9003 = 15.1 mL.

Again, note the flow of the calculation as shown by the arrows. After converting all quantities of reactants into grams by whatever means necessary, we go down the table and divide each reactant by its molecular weight. After selecting the limiting reactant and rejecting other reactants, we go up the table to calculate expected yield of product, multiplying the moles of limiting reactant by the molecular weight of the product.

In this reaction we use twice as many moles of ethyl alcohol as would be required to react with all of the acetic acid. We often use such an excess of one reactant, or an even greater excess if necessary, when the reaction is an equilibrium reaction with a rather unfavorable equilibrium constant. Sometimes we can drive the reaction to the right (toward the formation of product) by using an excess of one reactant.

For most of the syntheses in this text, the above calculations will suffice; however, we have made an assumption about the stoichiometry of the reaction that will not always be valid. More generally, the balanced reaction would be written as follows:

$$a\text{A} + b\text{B} \longrightarrow c\text{C} + d\text{D}$$

where a, b, c, and d are the numbers (or coefficients) needed to balance the equation. In our second example, all of these numbers were the same:

$$a = b = c = d = 1$$

Whenever this is true, the calculations may be carried out using the simple procedure based on moles of reactants given above. However, if *any* of these numbers is not equal to one, then we must change our calculation to reflect the stoichiometry. One simple way of doing this uses equivalents of reactants rather than moles of reactants. Be careful not to confuse the terms *equivalents* and *moles*. **Equivalents** refers to *both the amount of reactant used and the specific chemical equation that we write;* **moles** refers only to the amount of reactant. To see why this is so, we rewrite the equation slightly by putting parentheses around each coefficient and chemical symbol:

$$(a\text{A}) + (b\text{B}) \longrightarrow (c\text{C}) + (d\text{D})$$

The quantities in parentheses are the equivalents of the substance; that is, one equivalent of A = a moles of A, one equivalent of B = b moles of B, and so on. Since the values of a, b, c, and d depend on the exact form of our balanced equation, the equivalents of A, B, C, and D will also depend on the way we write the equation. We can interpret the equation as saying either (1) that *one* equivalent of A reacts with *one* equivalent of B to yield *one* equivalent of C and *one* equivalent of D or (2) that a moles of A reacts with b moles of B to yield c moles of C and d moles of D. This means that the following ratios are true for the balanced equation:

(equiv. of A) : (equiv. of B) : (equiv. of C) : (equiv. of D) = 1 : 1 : 1 : 1

and, if A, B, C, and D have the same molecular weights assumed above,

$$\frac{\text{weight A}}{a \times 60} = \frac{\text{weight B}}{b \times 80} = \frac{\text{weight C}}{c \times 100} = \frac{\text{weight D}}{d \times 40}$$

In our experiments, where the quantities of reactants may not be balanced and where one or more reagents may be present in excess, we can calculate the number of equivalents of A directly or in one step by dividing the weight of A by the product ($a \times$ the molecular weight of A) and similarly for B, C, and D, as shown below. However, it is still desirable to calculate the number of moles of reactants and products because the number of moles is an absolute value, independent of how we write the equation. Therefore, we usually calculate the number of equivalents of a reactant from the number of moles of that reactant *for a specific equation* as follows:

$$\text{Equivalents of A} = \frac{\text{Weight of A}}{a \times (\text{Molecular weight of A})} = \frac{\text{Moles of A}}{a}$$

$$\text{Equivalents of B} = \frac{\text{Weight of B}}{b \times (\text{Molecular weight of B})} = \frac{\text{Moles of B}}{b}$$

Now our limiting reactant is the one whose number of equivalents is the smaller. Our calculation of the theoretical yield must also be changed to reflect the more complex stoichiometry and may be different for each of the products. Let us assume that we are trying to prepare product D; that is, C is a coproduct (*not* byproduct) of lesser interest. Then the calculation of the theoretical yield of D will be:

Theoretical yield of D = $d \times$ (Equivalents of limiting reactant)

$\times$ (MW of product)

The percentage yield of D is calculated exactly as shown in the earlier examples since the percentage yield depends only on the actual and theoretical yields.

Next we consider a specific example and a format for writing such calculations that may simplify the calculations. In this experiment we cause 15.0 g of succinic acid to react with 12.0 mL of ethyl alcohol and expect diethyl succinate and water as our products, as shown in the balanced equation. We obtain the molecular weights of the reactants and expected products from a chemical handbook if we can, or calculate them from the atomic weights if we cannot. We obtain the density of ethyl alcohol from the handbook. Then we write the equation, the quantities of reactants, molecular weights, equivalents, theoretical quantities of products, etc., in the following format.

$$
\begin{array}{c}
CH_2CO_2H \\
| \\
CH_2CO_2H
\end{array}
\;+\; 2\,CH_3CH_2OH \;\longrightarrow\;
\begin{array}{c}
CH_2CO_2CH_2CH_3 \\
| \\
CH_2CO_2CH_2CH_3
\end{array}
\;+\; 2\,H_2O
$$

	$C_4H_6O_4$	C_2H_6O	$C_6H_{14}O_4$	H_2O
	Succinic acid	Ethyl alcohol	Diethyl succinate	

Quantity		15.0 g ③	9.5 g (12.0 mL)	⑧ ↑	17.5 g ⑨ ↑	3.7 g		
MW		118.09	46.07		174.2	18.015		
Moles	①	0.127 ④	0.206	⑦	0.103	0.206		
Equivalents	②	0.127 ⑤	0.103	⑥	0.103 ⑥	0.103		

→×

Density   0.7893

The calculated numbers are obtained using the equations given above in a stepwise fashion:

Step 1   Moles of succinic acid = 15.0/118.09 = 0.127.

Step 2   Since there is an implied "1" in front of the molecular formula for succinic acid, equivalents of succinic acid = 0.127/1 = 0.127.

Step 3   The quantity of ethyl alcohol used is given in milliliters; therefore, the weight of ethyl alcohol = 12.0 × 0.7893 = 9.5 g.

Step 4   Moles of ethyl alcohol = 9.5/46.07 = 0.206.

Step 5   Since our balanced equation specifies two ethyl alcohol molecules, equivalents of ethyl alcohol = 0.206/2 = 0.103.

Step 6   The smallest equivalents of reactant is seen to be that of ethyl alcohol; therefore, ethyl alcohol is the limiting reactant. Thus, we expect only 0.103 equivalents of each product. We enter 0.103 in each product column.

Step 7 Since there is an implied "1" in front of the molecular formula for diethyl succinate, the expected moles of this product = $1 \times 0.103 = 0.103$.

Step 8 The theoretical yield = $0.103 \times 174.2 = 17.5$.

Step 9 Here we are not particularly interested in calculating the amount of water formed, but we do so for completeness.

Note again the flow of the calculation as indicated by the arrows. After converting all quantities of reactants into grams by whatever means necessary, we go down the table and divide first by the molecular weight, then by the coefficient (number) in the balanced equation. After selecting the limiting reactant and rejecting other reactants, we go up the table to calculate expected values, multiplying first by the coefficient in the equation and then by the molecular weight of the product.

Next, we assume that the student running this reaction obtained 13.8 g of purified product. With this information we make the final calculation.

Step 10 The percentage yield = $(13.8/17.5) \times 100 = 79\%$.

Some instructors prefer not to include the calculated values in the table of reactants and products but to report these in a separate section following the table. Some prefer to have all of the physical constants of the reactants and products included in the table, as we have shown for the density of ethanol; others prefer a separate table for these values. We have every confidence that your instructor will inform you of his or her preferences.

1. Some articles of commerce can be counted directly or indirectly. Some examples of articles that can be counted by weighing include coins of a given denomination, nuts, bolts, screws, marbles, and golf balls. Let us assume that we are making fasteners from nuts and bolts, that is,

$$1 \text{ Nut} + 1 \text{ Bolt} \longrightarrow 1 \text{ Fastener}$$

We find that 100 nuts weigh 120 g, and 100 bolts weigh 200 g. (a) Calculate the theoretical yield of fasteners if we buy one pound each of nuts and bolts at the hardware store. (b) Suppose that we find that two nuts are required on each bolt for a truly secure fastener. Calculate the theoretical yield of fasteners having two nuts.

2. For the example involving succinic acid and ethyl alcohol, rewrite the balanced equations as follows and recalculate the results.

$$\tfrac{1}{2} \text{ Succinic acid} + 1 \text{ Ethyl alcohol} \longrightarrow \tfrac{1}{2} \text{ Diethyl succinate}$$

Which, if any, numbers are changed in the recalculation?

3. For the following reaction calculate the theoretical and percentage yields of the organic product.

$$C_6H_{12}O \longrightarrow C_6H_{10} + H_2O$$

Quantity	10.0 mL	
Density	0.9624	0.8102

The student running the reaction obtained 5.0 mL of product.

4. For the following reaction calculate the theoretical and percentage yields of the organic product, $C_7H_{16}O$. Note that in some reactions involving inorganic substances we may abbreviate the equation in such a way as to leave out some simple and fairly obvious steps to emphasize the organic compounds in which we are interested. Your instructor and your increasing experience will tell you when this can be done. Here we show an intermediate acidification step that removes inorganic salts by putting H_3O^+ under the arrow. Although the intermediate step has been eliminated to give the abbreviated equation, this equation is (and must be) balanced with respect to starting materials and final products.

$$2\ C_3H_7MgBr + C_2H_4O_2 \xrightarrow[\text{then } H_3O^+]{} C_7H_{16}O + CH_4O$$

Quantity	20 mL of 3M solution	1.50 mL
Density		0.9742	0.8183

The student running the reaction obtained 1.8 mL of product.

5. When we plan a new experiment, one not found in a manual, we usually begin with the more valuable reactant and calculate the moles or equivalents of it. Using that quantity we calculate the amounts of other, less valuable reactants to use. In the third example given above both succinic acid and ethyl alcohol are

relatively inexpensive; however, other considerations suggest that succinic acid should be the limiting reactant and that ethyl alcohol should be present in excess. Revise the method of calculation and calculate how much ethyl alcohol we should use in order to have a 50% excess of it.

6. Three competitive laboratories simultaneously report new syntheses of the important pharmaceutical product, Curitall. Laboratory A reports a seven-step synthesis with yields in steps one through seven of 92, 91, 90, 85, 85, 80, and 70%. Laboratory B reports a seven-step synthesis with yields of 70, 80, 85, 85, 90, 91, and 92%. Laboratory C reports a four-step synthesis with yields of 74% in each step. Calculate the overall percentage yield for each laboratory. Assuming that the only differences in the three syntheses are the yields obtained, rank the three syntheses and explain your reasoning for assigning the relative ranks.

7. If you have access to a personal computer, you may wish to write a program that will accept as input from the terminal the information necessary for the calculation either in the form of a typed-in balanced equation or as answers to prompting questions plus information about the quantity of each reactant in grams or milliliters of liquid or solution, as well as the density or molarity of liquids or solutions. The outputs would be the moles and equivalents of both reactants and products and the theoretical yields of all products. This should be a relatively simple project in a modern computer language such as one of the dialects of UCSD Pascal or Modula-2 and only moderately more difficult in BASIC or FORTRAN.

Laboratory Reports

Forms for reporting results appear at the end of each experiment. Each report form provides space for the essential information needed to describe the experiment and to evaluate results. There are blanks for reaction equations, quantities required (expressed in grams and moles), physical constants, percentage yields, results of tests, etc. Data should be recorded as the experiment is performed—not hours after it is completed. The report forms are detachable and provide a convenient and uniform record of work done. However, should the instructor prefer that students keep a laboratory notebook, the report forms may be detached and secured with a strip of transparent tape to the right-hand pages of a hardcover, bound notebook, reserving the left-hand pages for observations, comments regarding the experiment, and answers to the questions that follow each report. Do all reporting in ink.

Certain synthesis experiments require that you submit the purified product to your instructor. Inasmuch as you have spent several hours preparing it, the fruit of your labor deserves to be placed in a clean, tightly stoppered vial with all pertinent information on an affixed, legible label. The label should include the name of the compound and structure (if space permits), the gross and tare weight of the vial, the net weight of the contents, observed bp or mp, your name, and date. A sample label is shown below.

For most of the laboratory courses for which this manual is intended, the use of the report forms as suggested above will suffice. However, in some courses the instructor may wish to require a more formal laboratory notebook, somewhat similar to the notebooks used in research laboratories. There is a variety of reasons for so doing. Your instructor may want you to learn to write proper reports as a part of a university- or college-wide emphasis on writing skills. Your instructor may want you to acquire good habits in the reporting of scientific results. Whatever the reasons, he or she will have to provide you with a description of the specific format required for the different types of experiments conducted in the organic chemistry laboratory: chemical tests and reactions, syntheses, and identification of unknowns, for example. This is necessary because there is no universally accepted format for laboratory notebooks, although there are widely accepted general guidelines. Thus, your instructor may modify or even replace the format for a notebook described here to meet the requirements of your course in organic chemistry.

Generally, a laboratory notebook should be a bound notebook, not spiral or loose-leaf. Each page should have ruled horizontal lines and one vertical ruled line not too close to the left margin. The pages should be numbered. In many laboratories the right-hand page is used for all entries pertaining to experimental procedures, observations, data, and results. The left-hand page is left blank or is restricted to such entries as your comments, excerpts from the chemical literature or handbooks, and comments by the instructor. Use the first page of the notebook as a title page that consists of the name and number of the course, your name, and an address or telephone number (not necessarily your own) to be used in the event that you lose the notebook. The next several pages may be left blank to be used as a table of contents, which must be kept current as you use the notebook. Each new experiment should begin on a new page, and it is generally considered poor practice to leave blank pages between experiments in research notebooks, even where the previous experiment is not yet completed. However, you may be allowed to do so for the experiments in this manual. Enter the dates of all experiments and measurements in the space between the left margin and the vertical ruled line. For some experiments it may be useful to enter the time of an observation also. Make corrections by carefully crossing out (not obliterating) erroneous observations, as history has shown that what may appear to be an error today could be the discovery of tomorrow. Your instructor's requirements will include a description of how much detail is expected of you in writing up each experiment. A general criterion is that your writeup must be sufficiently complete that another student could repeat your experiment using the information given in your notebook. This does not mean that you must copy each experiment from this manual or another printed source into your notebook, but it does mean that you must cite very clearly what you did. If you followed a published procedure, at a minimum you will state that you followed the procedure, cite an exact reference to the procedure, and describe in detail any deviations from it. Because this laboratory is intended as a learning experience, your instructor may require more than this minimum of detail and ask you to describe fully what you did and what you observed. This is good practice for the years ahead when you may not have a published procedure to follow.

In this manual the experiments fall into three broad classes: characterization, identification, and preparation. In the characterization experiments we examine and compare the reactions of various classes of organic compounds. In identification experiments we attempt to identify an unknown compound on the basis of chemical tests, preparation of derivatives, or spectroscopic analyses. In preparation experiments we synthesize an organic compound in one or more steps from some other organic starting material. Some experiments may involve two or even all three classes. Suggested laboratory notebook formats for the three classes follow.

Characterization Experiments

1. Experiment number and title.
2. Purpose. A brief but concise statement of the objectives of the experiment. A few sentences should suffice.

3. Applications and results. Repeat the following for each reaction or test investigated:
   a. Equation for reaction or test.
   b. Outline of procedure followed for reaction or test.
   c. Observations made. Cite any color changes, precipitates formed, or any other phenomena associated with the experiment. Chemistry is an observational science, and you must train yourself to become a careful and thorough observer. Spectra, chromatograms, etc., may be bound into the laboratory notebook if they are sufficiently small to do so. Otherwise, keep a separate loose-leaf notebook of these materials and enter a reference to the material in that notebook into the laboratory notebook. Do not forget to label and date each spectrum and chromatogram properly.
   d. Conclusions. Observations without conclusions are of little value. State concisely your conclusions based on your actual observations, not what the manual may lead you to think you should observe.
4. Answers to assigned exercises, if your instructor wants these in the notebook.

You may complete items 1, 2, and 3a before coming to the laboratory if your instructor so directs. This will require leaving sufficient blank space in the notebook for procedures, observations, and conclusions or use of a separate page for each reaction or test (a somewhat wasteful practice).

Identification Experiments

1, 2, 3. The same as for characterization experiments.
4. Identification of unknown. Summarize your conclusions from the reactions, tests, and spectra to support the structure you have assigned to the unknown.
5. Answers to assigned exercises.

Preparative Experiments

1. Experiment number and title.
2. Purpose.
3. Reaction equation or reaction sequence. Write balanced equation(s) for the principal reaction(s) involved in converting the starting material to the desired product.
4. Table(s) of reactants and products. Your instructor will determine the format used here. One format for such a table is given in the section above entitled *Calculation of Yields in Synthesis*. Enter into the table all of the reactants and the expected products. Generally, chemicals used in purification steps are not included. Typical entries for each compound include:
   a. Name and structure.
   b. Molecular weight.
   c. Weight in grams. If a liquid is measured by volume, state that fact and enter

both the volume and the weight calculated from the volume. If a solution of known concentration is used, state both the volume and the concentration.

d. Moles. Calculated as described above under *Calculation of Yields in Synthesis*.

e. Equivalents. Calculated as described above.

f. Physical properties as given in this manual, a chemical handbook, or the chemical literature—not values you have measured.

5. Outline the procedure followed, including purification steps. Normally the description of the procedure followed is entered *while* or immediately *after* the experiment is performed; however, your instructor may ask that you outline the procedure to be followed *before* coming to the laboratory. In that event, enter into your notebook any changes that you make in the proposed procedure as you perform the actual experiment.

6. Observations. Report any physical properties of the product or of intermediates that you determined in the course of the preparation, including melting and boiling points, color, phase (liquid, crystalline, amorphous, gum, etc.), odor (**Caution!** *Only if your instructor so directs.*), optical rotation, spectra, and chromatograms. Report your observed yield in grams and as the percentage of theoretical. Data such as spectra and chromatograms may be bound into the laboratory notebook if they are sufficiently small to do so. Otherwise, keep a separate loose-leaf notebook of these materials and enter a reference to the material in that notebook into the laboratory notebook. Do not forget to label and date each spectrum and chromatogram properly.

7. Conclusions and Comments. These may include comments regarding principles demonstrated, success or failure of the experiment, sources of error, and suggestions for improvement.

8. Answers to assigned exercises.

The format for a preparative experiment is illustrated below with a sample entry from the laboratory notebook of a typical student. Since the experiment being reported is not in this edition of this manual, the student had to describe the procedure in detail. Note also the student's suggestions at the end of the writeup.

EXPERIMENT 31: Preparation of N–tert–Butylacetamide

Purpose: To prepare an N–tert–butyl amide by the Ritter Reaction.

tert–butyl   acetonitrile   N–tert–butylacetamide
 alcohol

Quantity	16.0 g (22 mL)	4.10 g	10 g	11.5
MW	74.12	41.05	98	115.18
Moles	0.216	0.100	0.10	0.100
Equivalents	0.216	0.100		0.100
Density	0.7887(20/4)	0.7857(20)		
bp	82.2° (760)	81.6°		
mp	25.5°	−45.72°		97–98°

Procedure:

3 Nov 86   In a 125–mL Erlenmeyer flask were mixed 22 mL (16 g, 0.216 mole) of <u>tert</u>–butyl alcohol, 4.10 g of acetonitrile, and 50 mL of glacial acetic acid. Ten grams of concentrated sulfuric acid was added dropwise from a dropping funnel. The flask was swirled after the addition of each drop of acid to mix the reactants and to keep the temperature below 20°. After all of the sulfuric acid had been added, the flask was stoppered loosely and stored in the desk.

17 Nov 86   The reaction mixture was poured into 200 mL of water in a 600–mL beaker and was neutralized by slow addition of solid sodium carbonate. The organic layer was separated from the slightly alkaline aqueous layer (pH paper) using a separatory funnel. The aqueous layer was extracted with three 50–mL portions of ether, which were combined with the organic layer. The ether was removed by distillation using a warm water bath (50°) in the hood. The syrupy, residual liquid was pale amber in color and solidified on cooling. The crude crystals were collected on a Büchner funnel, then recrystallized from hexane.

Weight of dry product = 8.8 g
Theoretical yield   = 11.5 g
Percentage yield   = 77%

mp = 96–97°
lit mp = 97–98°

infrared spectrum IR–13

Comments: This procedure shows that a <u>tert</u>–butyl cation generated from the reaction of <u>tert</u>–butyl alcohol with sulfuric acid can be trapped with acetonitrile. The procedure worked very well, but some better way of adding the sulfuric acid is needed as handling it with a separatory funnel was awkward. Perhaps sulfuric acid could be added by careful use of a medicine dropper or diluted by carefully adding it to some of the acetic acid before transfer to the separatory funnel.

When writing in your notebook it is wise to use the currently accepted punctuation, abbreviations, and grammar. Some guidelines for current usage are given below in the form of examples.

Ten milliliters of acetic acid . . . *if a sentence begins with a number, spell out the number and any unit of measure.*

A 10-mL portion of acetic acid . . . *a way to avoid spelling the number, but note the hyphen that is required.*

To a 100-mL, 3-necked flask 30 mL of acetic acid was added . . . *note where the hyphens are and are not used.*

Note that, in the foregoing examples, the pronoun "I" never appears. Do not write "I added 10 g . . ." Write instead "Ten grams was added . . ."

Common abbreviations:

gram	g
milligram	mg
liter	L
milliliter	mL
second	sec
minute	min
hour	h
melting point	mp
boiling point	bp
mole	mol
equivalent	equiv

If you would like to learn more about the art of writing laboratory notebooks, the American Chemical Society has published a short book on the subject that should be useful to all science students: Howard M. Kanare, *Writing the Laboratory Notebook*, American Chemical Society, Washington, D.C., 1985.

Ground-Glass Equipment

Cork- and rubber stopper-type glassware once commonly used in beginning organic laboratories has been replaced almost entirely by tapered ground-glass, jointed glassware. This equipment, usually referred to as standard-taper ($) glassware, has a number of advantages. All pieces of equipment with ground-glass joints of one size are interchangeable. The need for boring stoppers is eliminated and apparatus can be assembled in far less time. Assemblies illustrated in this manual show standard-taper ware. However, the use of standard-taper equipment requires the exercise of special techniques. Such ware, when connected, makes a rigid joint, and an assembly must be properly supported to prevent any undue strain on the parts. Before assembling any apparatus with ground-glass joints, lightly grease the joint surfaces to make an airtight seal and to prevent the joint from "freezing." Then begin the assembly of the apparatus *from the bottom up*. A lower section should not be expected to hold fast to an upper section simply because the joint is greased and fits snugly. If the lower section is not supported, it may slip off and break. During assembly, a simple expedient for preventing accidental breakage is the use of rubber bands or long "baggie" ties to hold sections together temporarily. However, before the apparatus is used, all parts must be properly supported by clamps as indicated in this manual's diagrams. At the end of the experiment disassemble the apparatus as soon as it is safe to do so, again to prevent the joints from freezing. If, despite your precautions, the joints should freeze, ask your instructor for help disassembling them before attempting heroic measures on your own.

Methods of Heating

Many of the procedures used in the organic chemistry laboratory require the heating of a substance or a mixture of substances: carrying out a reaction, distillation of a liquid, recrystallization of a solid, determination of a melting or boiling point, etc. Naturally, we would like to have a method of heating that would be safe, easy to use, and inexpensive. Unfortunately, there is no one method that meets all of these criteria for every application. In your laboratory your instructor will have had to choose a method for each experiment in which heat is involved based on a compromise between the three criteria. A principal concern, regardless of the method used, is that it be used with the full understanding of the advantages and disadvantages of the method and the safety factors involved. Read and understand this section of the manual *before* doing any experiment requiring the application of heat, and refer to it from time to time to refresh your memory, for the generalizations given here apply to heating in any laboratory situation, not just the organic chemistry laboratory.

There are five common sources of heat in the laboratory: (1) Bunsen or other burners, (2) heated oil, water, or other baths, (3) electric heating mantles, (4) electric hot plates, and (5) steam baths. In this manual most of the illustrations show the use of Bunsen burners, in part because the use of burners is the most demanding of proper technique, and in part because many laboratories are not yet fully equipped with alternative heat sources. Also, all parts of the apparatus in the illustrations are more easily seen with no bath or mantle to obscure the view. However, unless stated otherwise, burners can usually be replaced with oil baths or with mantles, and in some instances with steam baths or hot plates. After the first few experiments we do not show a source of heat, and the student is expected to use the principles learned in the earlier experiments to provide a safe and efficient heat source based on the equipment available.

In general, the alternative methods of heating are used in many different ways in the laboratory—too many to describe in detail here. Therefore, for illustrative purposes, we have chosen one of the most common techniques in organic chemistry requiring the application of heat: the heating of a substance or mixture of substances "under reflux." Usually the apparatus used consists of a round-bottomed flask equipped with a water-cooled condenser, appropriately attached by clamps to a ring-stand. The mixture in the flask is heated to boiling, causing vapors of the liquid to rise into the condenser where they are condensed on the cool surface and returned to the flask. If the rate of heating is properly controlled, the mixture will *reflux* steadily and gently for a long time and little or no vapor of the boiling liquid will escape into the laboratory. If you understand the principles of heating this apparatus, you should be able to extend these to the other apparatus setups used in organic chemistry. Some of these other setups are more demanding than the simple reflux apparatus; therefore, additional information will be provided as required. In particular, the additional heating requirements for distillation are discussed in Chapter 3.

1. Heating with Bunsen or Other Burners

The setup used for heating a reflux apparatus with a Bunsen burner is shown in Figure 0.1, which is largely self-explanatory. Note the positions of clamps, direction of

Figure 0.1   Apparatus assembly for refluxing liquids using a burner.

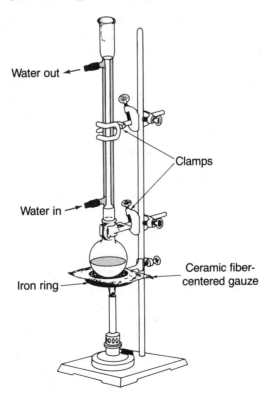

water flow, and all the other detailed information provided. The rate of heating is controlled by the size of the flame and by the distance of the flame from the flask. A ceramic fiber-centered wire gauze supported on an iron ring is used between the flame and the boiling flask to prevent the formation of "hot-spots" or uneven heating at the point where the flame touches the flask. Uneven heating can cause vigorous or violent local evolution of vapor or "bumping," which is alleviated in part by the wire gauze and in part by placing a "boiling stone" in the flask *before* you begin heating.

Caution!   *Recent studies have indicated that using wire gauzes without some heat-resistant center can cause severe local hot-spots and can lead to cracking of the flask. The asbestos-centered wire gauze used for so many years in organic chemistry laboratories should no longer be used because of the known carcinogenic properties of asbestos fibers.*

The advantages of heating with a burner are: (1) the apparatus is easy to assemble, disassemble, and clean, (2) the flask can be heated rapidly or cooled down rapidly, and (3) the cost is relatively low. The disadvantages are: (1) the danger of fire is greatest with this type of heat, and (2) even heating of the flask is difficult. If a burner is used for heating, do not leave it unattended as the flame could go out, leading possibly to the accumulation of enough gas to cause an explosion. Burners are especially hazardous when used with low-boiling, flammable liquids. If they must be used with such liquids, carefully assemble the apparatus with sound glassware and properly greased joints and use it with diligence and care. If the flask is heated too strongly, vapors of the flammable materials could escape from the condenser and be ignited. If any part of the apparatus should crack while being heated with a burner, a fire is likely to result. Also, before lighting any burner you must check to see that no flammable liquids are being used or transferred by other students working near you.

Caution!   *Do not use burners to heat any flammable liquid in an open vessel.*

In general, the only liquids heated in an open vessel by a burner are those used in water or oil baths or in melting point baths.

2.   Heating with Oil, Water, or Other Baths

One setup used for heating a reflux apparatus with an oil bath is shown in Figure 0.2. In this method of heating the mixture in the flask is heated by the hot liquid in the bath, which is itself heated by (1) a Bunsen burner, (2) an electrical heating element immersed in the bath, or (3) an electric hot plate. If a burner is used, control the rate of heating by adjusting the flame, and observe all of the precautions cited in section 1 above. If an electrical heating element is used, it will be attached to some device, often a variable transformer (Variac or Powerstat), that controls the heat the element puts out by controlling the voltage to the element. *Never attach the element directly to an electrical outlet!* To do so would very probably burn out the heating element and possibly expose you to electrical shock. If a hot plate is used, control the rate

Figure 0.2   Apparatus assembly for refluxing liquids using an oil bath.

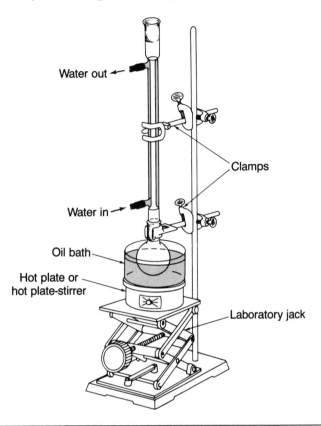

of heating by adjusting the temperature control of the hot plate. In Figure 0.2 only the oil bath heated by a hot plate is shown, as this is probably the most reasonable alternative for an elementary organic chemistry laboratory. In a more advanced laboratory, the use of an electrical heating element may be more advantageous.

The advantages of heating with an oil bath are: (1) the temperature of the bath can be measured with a thermometer and easily controlled by adjusting the heat source, and (2) heating is uniform around that part of the flask immersed in the bath. Disadvantages include: (1) the maximum temperature that can be used safely is limited by the properties of the oil or other bath liquid, (2) oil baths heat up and cool down slowly, requiring in the latter instance some means of quickly removing the bath from the flask or vice versa, (3) storage of the bath is inconvenient, and (4) cleaning apparatus immersed in oil can be a nuisance.

The choice of the liquid to be used in the bath is based on heating requirements, ease of use, and expense. Mineral oil baths are relatively inexpensive but should not be heated above 200°C as they begin to smoke at about that temperature and are, especially if reused frequently, susceptible to flash ignition of the vapors at higher temperatures. Silicone oil baths are more expensive; however, they can be heated safely somewhat above 200°C before decomposition begins. For the elementary organic laboratory the use of polyalkylene glycols such as Carbowax 600 or UCON Heat Transfer Fluid 500 is an excellent compromise as the material is relatively inexpensive, stable up to temperatures of about 160°, and water-soluble. The latter is an especially desirable trait in the elementary laboratory as cleaning flasks immersed in mineral oil—and even more so in silicone oil—requires the use of organic solvents and the facilities for disposing of waste solvents. All oil baths are susceptible to traces of water. Water in mineral oil or silicone oil baths can cause spattering of the hot oil, and any bath containing droplets of water should not be used. Water in polyalkylene glycol, which is somewhat hygroscopic, may temporarily lower the temperature reached by the oil until the water boils out.

Caution! *Any movement of a bath filled with hot liquid is hazardous; do it only if absolutely necessary. Before doing so, turn off the electrical power to the heat source, preferably by turning off the switch first, then unplugging the cord. Oil baths retain heat for some time and should be allowed to cool before being handled. If they must be moved while hot, exercise great care.*

When oil baths are used for heating, make some provision for removing the source of heat quickly if necessary. In an emergency it may not be possible to move the glass apparatus; thus, the bath may have to be moved. Probably the best (and most expensive) way of moving the bath vertically involves the use of a laboratory jack as shown in Figure 0.2. A less expensive alternative is to use a bath with a handle (such as a porcelain casserole or an enameled steel pan) so that you can support the bath with one hand while removing the hot plate, if used, with the other. If the bath is heated by an electrical heating element and no jacks are available, a supply of scrap lumber blocks somewhat larger in area than the item to be supported and of varying thicknesses should be provided. You can stack these blocks under the bath to elevate it into position, provided that the stack is not so tall that it is unstable.

A very useful heated bath for low-boiling organic liquids is the hot-water bath. Such a bath can be heated by a burner, provided that the burner is turned off whenever flammable liquids or vapor could come into contact with the flame. Use of a hot-water bath heated by a hot plate is generally very safe; however, some hot plates are not explosion-proof; that is, they may utilize relay contacts that can spark when the current is turned on or off by the temperature control system.

Caution! *Do not heat flammable liquids in an open vessel in any bath that is being heated by a hot plate.*

Heat any flammable liquids in an open vessel only in a good fume hood, even if there are no flames or operating hot plates nearby.

For the removal of low-boiling solvents by distillation, a hot-water bath may be used without a direct heat source. Hot water can be poured into the bath (preferably a metal bath or pan) from an ordinary kettle. The kettle may be heated in a hood set aside for this purpose or in some remote location. Our experience has been that one filling of hot water is enough to distill 100 mL of ether, the solvent most commonly removed using a hot-water bath.

3. Heating with Electric Heating Mantles

One setup used for heating a reflux apparatus with an electric heating mantle is shown in Figure 0.3. In this method of heating, the mixture in the flask is heated by a resistance element embedded in a mat of glass fiber encased in a glass-fabric cover. The power cord leading to the resistance element of the mantle is attached to a variable transformer (Variac or Powerstat). *Do not attach the mantle directly to an electrical outlet, as few mantles can survive the full line voltage for any length of time*. To do so runs the risk not only of destroying the mantle but also of exposing yourself to electric shock. Control the temperature of the mantle by varying the voltage applied with the variable transformer. Most mantles are supplied with a built-in thermocouple, as evidenced by two small wires coming out of the glass fabric. With an appropriate meter the thermocouple can be used to monitor the

---

Figure 0.3   Apparatus assembly for refluxing liquids using a heating mantle.

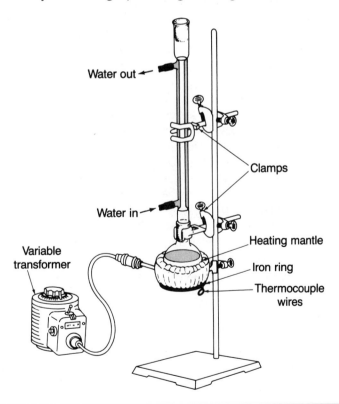

temperature of the mantle. However, because of the expense of the meters, this is rarely done in the organic laboratory. Note that the heating mantle must be supported either on an iron ring, on a laboratory jack, or on a low stack of blocks so that the mantle can be removed quickly, *after turning off the power!*

The advantages of heating mantles are: (1) fire hazards are less, but not entirely absent, (2) mantles are easy and clean to use, and (3) often apparatus heated by mantles can be left for long periods of time without close supervision. Disadvantages include: (1) heating mantles are not only expensive but also must be available for every different size of flask expected to be used, (2) mantles heat up and cool down slowly, (3) mantles are susceptible to damage from spilled chemicals, and (4) if the level of the liquid in the flask is below the level of the mantle, that part of the flask that is heated by the mantle but not cooled by the liquid may develop local hot-spots resulting in decomposition, bumping, or even destruction of the mantle.

Use heating mantles only with the size of flask for which they are intended. Packing glass wool around a small flask in a larger mantle is usually bad practice. Heating mantles work best when the flask is filled somewhat above the level of the mantle and the mixture is being heated under reflux. When a mantle is being used to heat a distillation apparatus, turn off the electric power before the flask is nearly dry or the mantle may burn out. As noted above, the mantle must be supported above the desk top in some manner such that it can be quickly removed if necessary. In addition, cracking of desk tops has been observed when large mantles have been left unattended on the desk. In some experiments it may be desirable to keep a cold-water or other cooling bath nearby in the event that the flask needs to be cooled quickly after removal of the mantle. Do not spill water or other chemicals on the fabric of the mantle.

An alternative to the heating mantle is the Thermowell round-bottomed flask heater (Fig. 0.4a). In appearance, this device resembles a heating mantle in a metal container; however, there are important differences between the two devices. The Thermowell heater consists of a resistance element embedded in a high-alumina refractory heating shell supported by other refractory materials and contained in a steel body with an aluminum bottom (to permit the use of magnetic stirrers). Generally, a more sophisticated temperature controller is required. These heaters have the advantages of the heating mantle with fewer disadvantages. Their construction makes the *devices* less susceptible to overheating, and flasks smaller than the nominal size of the heater may *sometimes* be used. However, the organic substances being heated are still susceptible to overheating. Thus, low-boiling flammable liquids should *not* be heated in flasks smaller than the design capacity of the heater; nor should liquids be distilled to near dryness in these heaters. Avoid spillage into these heaters as with heating mantles.

4.   Heating with Electric Hot Plates

Two types of hot plates are commonly available in the organic chemistry laboratory, those with (Fig. 0.4c) and without (Fig. 0.4d) built-in magnetic stirrers. The former

Figure 0.4   Heating devices. (a) Thermowell heater. (b) Steam bath. (c) Hot plate-magnetic stirrer. (d) Hot plate.

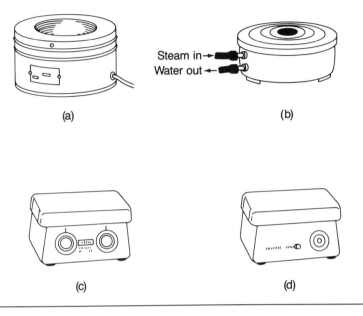

Steam in →
Water out ←

(a)   (b)

(c)   (d)

provide the convenience of a magnetic stirrer and hot plate in one device, albeit at a premium price. Simple hot plates and magnetic stirrers are often similar in appearance; thus, it pays to be sure that the proper device is being used. Hot plates are generally used to heat *flat-bottomed* containers such as beakers, Erlenmeyer flasks, casseroles, and oil or water baths. If a round-bottomed vessel is to be heated, an oil or water bath must be used in conjunction with the hot plate. The surface of the hot plate is heated by a resistance element similar to that used in heating mantles; however, the hot plate comes with a built-in control unit and should not be attached to a variable transformer. When hot plates are used to evaporate *nonflammable* liquids from an open vessel, the evaporation must be done in a good fume hood. Many, if not most, hot plates contain devices to switch the electrical current on and off. Some of these devices can generate electrical sparks; therefore, do not depend upon hot plates to be explosion-proof.

Caution!   *Do not use hot plates to evaporate flammable liquids from open vessels, even in a fume hood.*

If a hot plate must be moved up or down, support with a laboratory jack is highly desirable. The surface of the hot plate can be *very* hot; thus, handle the device with due care.

5. Heating with Steam Baths

For those laboratories that are equipped with piped-in steam, the use of steam baths or steam cones provides a safe and convenient means of working with many low-boiling, flammable or inflammable liquids or reaction mixtures. A typical steam bath is illustrated in Figure 0.4b. The bath has an inlet at the top for steam and an outlet at the bottom for removal of condensed water. The upper surface consists of a series of interlocking, concentric rings that may be removed or replaced to accommodate flat- or round-bottomed vessels of various sizes. When the opening has been adjusted, the steam valve is turned on *cautiously* and the accumulated water in the steam pipe is allowed to drain through the lower opening. When steam begins to issue from the opening, the flow of steam is controlled with the steam valve to achieve the desired rate of heating.

The advantages of the steam bath are: (1) much lower incidence of fire hazard, (2) rapid heating, and (3) low long-term cost if steam is available. Disadvantages include: (1) the maximum temperature achieved is about 90° and (2) water may condense on the apparatus, possibly contaminating the material being heated. Keep the rate of steam flow to the minimum needed to avoid excessive condensation. Although there are no experiments in this manual requiring the use of carbon disulfide, you should know that carbon disulfide can be ignited at steam bath temperatures and is one of those compounds that cannot be heated safely, even with steam.

Melting Points

The identification of a new organic compound is often a laborious and exacting task. A known compound, on the other hand, usually may be identified (or characterized) by the determination of one or more readily measurable *physical properties* (melting point, boiling point, refractive index, infrared, ultraviolet, or nuclear magnetic resonance spectrum) in conjunction with the examination of a few *chemical properties*. For solid substances one of the most useful physical properties is the melting point. The physical chemist defines the melting point of a substance as the temperature at which the solid and liquid phases are in equilibrium. The determination of such melting points as a routine identification procedure is difficult, especially with a limited amount of material. Fortunately, such melting points are unnecessary in the work of the organic chemist. The organic chemist utilizes instead a **capillary melting point.** The capillary melting point is a *range* of temperature over which a minute amount of the solid in a thin-walled capillary tube first visibly softens and finally completely liquefies. For the sake of simplicity, such melting point ranges are called **melting points** (mp), and organic chemists have used them for a century or more. These melting points are just as meaningful to the organic chemist as the melting points of the physical chemist, and all of the recorded melting points in handbooks, journals, and other organic chemical literature, unless otherwise specified, are of this type.

The melting point range of a solid organic compound, if very nearly pure, will be small (0.5–1.0°), and the substance is said to melt **sharply.** The presence of impurities, even in minute amounts, usually depresses, or lowers, the melting point and widens the range. Thus, the melting point not only is useful as a means of identification but also as a criterion of purity. Depression of the melting point of a pure compound by an impurity also is useful as an aid in identification. Consider, for example, a substance X thought to be identical with the compound Y because X and Y have the same melting point. If we mix a small amount of X and Y together and measure the melting point of the mixture, we determine what is called the **mixture melting point** of X with Y. If the mixture melting point is the same as the melting point of Y, X and Y *probably* are identical. (Exceptions to this generalization are relatively uncommon but are known.) However, if the mixture melting point is depressed below the melting point of Y (and the range broadened), then X acts as an impurity in Y and the two cannot be identical.

In this experiment you will determine (a) the melting points, separately, of a pair of pure organic compounds that melt at or near the same temperature, (b) the mixture melting points of the same pair, and (c) the melting point of an unknown compound.

PROCEDURE   A. Melting Point Apparatus

A laboratory melting point apparatus should meet several design objectives. First there must be a convenient source of heat, usually a gas flame or an electrical resistance heating element. Second, the temperature rise in the vicinity of the sample must be controllable. With a gas flame or an electrical element this may be accomplished by using a reasonably large volume of some thermally stable liquid, usually mineral oil, silicone oil, or polyalkylene glycol (Fig. 1.1 and Fig. 1.2), so that most of the heat being applied is expended in heating the oil rather than the very small sample. Electrical devices may employ a large block of metal as a heat sink or a

---

Figure 1.1   The melting point bath and capillary. The stirrer is moved slowly up and down to ensure uniform heating of the oil.

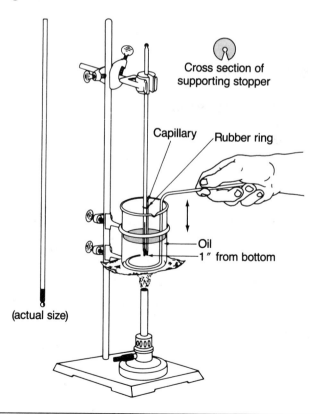

Figure 1.2   The Thiele melting point apparatus.

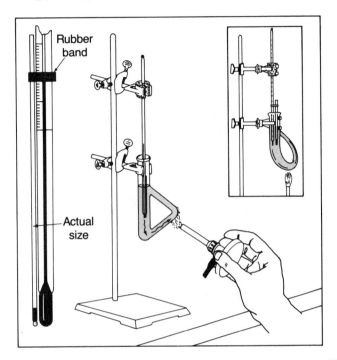

carefully designed low-wattage heating element and a smaller block (see Fig. 1.4 on page 38). Third, the temperature in a reasonably large zone in the vicinity of the sample must be as nearly uniform as possible so that both the sample and the temperature measuring device are at the same temperature. This is usually accomplished by manually or mechanically stirring the oil (Fig. 1.1), by convection currents in the oil (Fig. 1.2), or, in the case of a sophisticated electrical device, by careful engineering and, possibly, insulating the heat sink or block (Fig. 1.4). Finally, some means of measuring the temperature must be provided. Although this can be as elaborate as an expensive themistor device, in all of the apparatus setups described here a laboratory thermometer is used. The thermometer must be oriented so that its bulb is as close to the sample as possible (i.e., in the zone of uniform temperature). In careful work the thermometer will be "calibrated" *in the apparatus* by determining the melting points of a series of *pure* compounds whose melting points cover the range of the thermometer.

All of the apparatus setups described here will give satisfactory melting points. They are presented in order of increasing complexity and cost. The electrically heated device described in Procedure A(c) is only one of several well-designed commercial melting point devices currently available. We have described it because students in our laboratories have indicated a preference for it over other devices made available to them.

Figure 1.3 Heating curves of Mel-Temp at different fixed voltages.

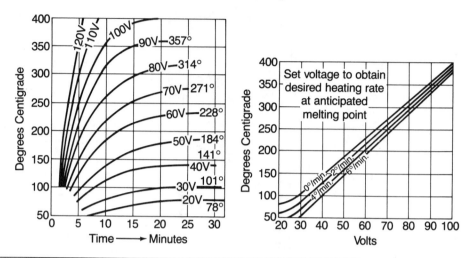

(a) Laboratory Constructed Apparatus

Assemble an apparatus like that shown in Figure 1.1. Support a 150-mL beaker on a ring-stand and fill it about two-thirds full of Nujol or some other high-boiling mineral oil or a polyalkylene glycol such as Ucon oil 500 or Carbowax 600. Fasten a ring around the beaker to prevent any accidental spilling of hot oil. Carefully bore a hole in a cork stopper large enough to accommodate a 260°C thermometer. With a sharp knife or razor blade cut out a one-sixth section of the cork and slide it over the thermometer in such a way that the calibrations are visible through the cut-out opening. Fasten the cork and thermometer in a buret clamp centered about six inches above the beaker. Lower the thermometer into the melting point bath by sliding it through the cork until it is about one inch from the bottom of the beaker. The buret clamp, when closed, will compress the cut cork and grip the thermometer quite firmly. Prepare a stirrer from a piece of solid glass rod 35 cm in length by heating and shaping one end into a ring. Bend the other end into a handle. Alternatively, you may construct a stirrer from a reasonably stout piece of nichrome wire of the same dimensions. In either case the ring at the end of the stirrer should be sufficiently large to clear the thermometer but not fit too closely to the wall of the beaker. Although this apparatus is shown being supported on a ceramic fiber-centered wire gauze (**Caution!** *Not asbestos-centered.*) and heated by a burner, heating by means of an electric hot plate is a safer alternative; however, control of the rate of heating with a hot plate is more difficult. If you use this apparatus, limit melting point determinations to those below 160° for polyalkylene glycol baths, below 200° for mineral oil baths, and below 250° for silicone oil baths.

(b) Thiele Tube

If Thiele melting point tubes are provided, assemble a melting point apparatus like that illustrated in Figure 1.2.

Support a Thiele tube on a ring-stand, using a buret clamp, and fill the tube to the top of the upper arm with a clear bath oil. Carefully bore a hole in a cork large enough to accommodate a 260°C thermometer. With a sharp knife or razor blade cut out a one-sixth section of the cork and slide it over the thermometer in such a way that the calibrations are visible through the cut out opening. Fasten the cork and thermometer in a buret clamp centered about six inches above the Thiele tube. Lower the thermometer into the Thiele tube by sliding it through the cork until it is about one inch from the bottom of the straight portion of the tube. The buret clamp, when closed, will compress the cut cork and grip the thermometer quite firmly.

Note that, as shown in Figure 1.2, Thiele tubes come in two different models, either of which will work satisfactorily if properly used. Both should be filled just above the level of the side-arm to allow for expansion during heating. Note also the part of the Thiele tube that must be heated with the burner if the tube is to function properly. The temperature limits for Thiele tubes are the same as those given in section A(a).

(c) Commercial Electrically Heated Apparatus

The Mel-Temp® capillary melting point apparatus (Note 1)[1] is illustrated in Figure 1.4. This device has a sealed electrical heating element, which provides a zone of uniform temperature by means of a high thermal conductivity block, heats each capillary on three sides, and accepts up to three capillary melting point tubes at one time. The melting point tube is illuminated by a built-in light source and is observed through a six-power lens system. To use the apparatus, insert a melting point tube in one of the capillary wells, set the voltage of the heating element to obtain the desired heating rate at the anticipated melting point using the voltage control and temperature vs. time or voltage curves (Fig. 1.3), and observe the melting point through the viewing lens. ***Caution!*** *Be sure to turn off the apparatus after each use to avoid serious damage to the heating element.*

B. Melting Points of Pure Compounds

Obtain six melting point capillaries from the storeroom or your instructor. If the capillaries are not sealed on one end, seal one end of each capillary by slowly rotating the end in the edge of a hot flame until a small bead appears (see Fig. 1.1). When cool, these capillaries are ready for filling. Have your instructor check your melting point apparatus and your capillary tubes before proceeding further.

---

[1]This reference is to notes usually found at the end of the procedure.

Figure 1.4 Commercial electrically heated capillary melting point apparatus.

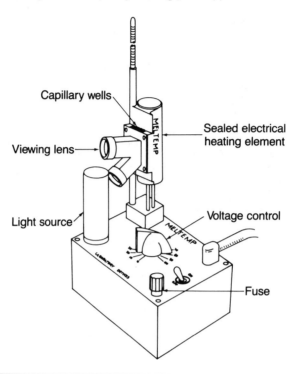

Place a small sample of any one of the twenty-five compounds listed in Table 1.1 on a piece of clean paper or clay plate. Crush the material to a fine powder with the spatula and scrape it into a small mound. Fill one of the melting point tubes by pushing the open end into the mound of powder, using the spatula as a backstop. When a small plug of powder has collected in the opening of the capillary tube, work the material down to the sealed end by scratching the capillary with a file while holding it lightly at the top or by very carefully tapping the closed end of the capillary on the desk surface.

Caution! *Hold the capillary loosely in your fingers if you tap it on the desk. If you hold it tightly, it might break with the risk of cutting you, possibly getting some of the sample in the wound.*

Repeat this process until a column of powder about 3–4 mm in height has collected in the capillary tube. Tamp the powder compactly in the capillary by dropping it on the desk several times through a two-foot length of glass tubing. The compacted powder should be about 2–3 mm in height.

If you are using an apparatus like that in Figure 1.1 or Figure 1.2, follow the procedure given in the next three paragraphs. If you are using the electrically heated

Table 1.1   *Melting Points of Some Common Organic Compounds*

Compound	mp, °C
Pimelic acid	103–105
Catechol	104–106
Azelaic acid	105–106
Resorcinol	109–110
Acetanilide	113–114
dl-Mandelic acid	117–118
Succinic anhydride	118–120
Benzoic acid	121–122
2-Naphthol	121–122
Urea	132–133
trans-Cinnamic acid	132–133
Benzamide	132–133
Acetophenetidine	134–136
Maleic acid	134–136
Malonic acid	135–137
Benzoin	136–137
Anthranilic acid	145–147
Cholesterol	148–150
Adipic acid	152–153
Citric acid	153–155
Salicylic acid	156–158
Benzanilide	160–161
Itaconic acid	163–165
Sulfanilamide	164–166
Triphenylcarbinol	164–165

device shown in Figure 1.4, *read* these paragraphs and modify the procedure as directed in the final paragraph of this section.

Attach the capillary tube to the thermometer with a small rubber band sliced from the end of a piece of rubber tubing. Position the capillary on the thermometer so that the filled portion is just opposite the mercury bulb (Fig. 1.2). Place the thermometer in the bath as before. Make sure that the small rubber band is well above the surface of the oil or the rubber will soften and disintegrate during the determination.

Heat the oil with a small flame and allow the temperature to rise fairly rapidly to within 15–20° below the melting point of the compound; then the burner should be adjusted so that the temperature rises no more rapidly than 2–3° per minute during the actual determination of the melting point. Watch the sample closely and check the temperature reading frequently. Write down the temperature at which the first visible softening of the sample is noted. Continue heating at 2–3° per minute and note the temperature at which all of the material has turned to liquid. The two values determined define the melting range or the melting point. Record your results.

Repeat the above process using a different compound with a melting point at or near that of the first. Discard melting tubes after one use and do not attempt to redetermine a melting point on a sample previously melted.

If you are using the electrically heated apparatus of Figure 1.4, modify the above directions as follows. Rather than attach the capillary melting point tube to the thermometer, insert the melting point tube in one of the three capillary wells. Rotate the voltage control knob to "0" and turn on the apparatus. Make certain that the sample in the melting point tube is visible in the viewing lens. Adjust the voltage control knob to a setting that will give a reasonably rapid rise (about five minutes) of temperature to within about 15–20° below the anticipated melting point. (To determine the voltage setting to be employed, use the family of curves in Figure 1.3 in which temperature is plotted against *time*.) When this temperature has been reached, quickly lower the voltage setting to a value that will give a heating rate of 2–4° rise per minute during the actual determination of the melting point. (Obtain the voltage setting to be used from the family of curves in which temperature is plotted against *voltage*.) You should be able to get satisfactory results by using an initial voltage setting of 60 volts; then at a temperature of 15–20° below the anticipated melting point quickly lower the voltage to 35 volts if your sample should melt between 100 and 120°, to 40 volts if it should melt between 120 and 140°, or to 45 volts if it should melt between 140 and 160°. The observation and recording of the melting point is carried out exactly as described above for melting points determined in an oil bath. After you have acquired confidence in your ability to use the instrument and interpret the heating curves, you may use it to determine as many as three melting points at once of different substances, to compare the melting points of different samples of the same substance, or to make mixture melting point determinations (as described below in Procedure C). Be sure to turn the voltage control to "0" and turn off the apparatus as soon as you have finished.

C. Mixture Melting Points

Take approximately equal amounts of the two compounds selected from Table 1.1 and mix thoroughly on a clay plate or a clean piece of paper. Better still, grind them into an intimate mixture with a mortar and pestle. Determine the melting point of the mixture and note carefully the effect one compound has upon the melting point of the other.

D. Determination of an Unknown

Obtain an unknown compound (it will be one listed in Table 1.1) from your instructor, record its number, and determine its melting point in the following way. Fill two melting point tubes with the unknown. Determine an approximate melting point using one tube and a rate of heating of about 15–20° rise per minute. If you are using the apparatus shown in Figure 1.4 for this experiment, a voltage setting of 65 volts should suffice. As soon as the sample melts, remove the burner and allow the bath to cool to 20° or more below the approximate melting point. Then quickly attach the second capillary tube to the thermometer and determine an accurate melting point using the approximate melting point as a guide in the determination. This procedure should be followed with all samples for which the melting point is not known in advance.

Identify your unknown from its melting point and from its mixture melting points with an identical compound from the side shelf (Table 1.1).

Cleanup: Dispose of all unused samples as your instructor directs. Do *not* return the chemicals to the original containers, as they may have become contaminated by handling. Discard used melting point tubes as your instructor directs.

Note 1 The Mel-Temp capillary melting point apparatus is available from Laboratory Devices, P.O. Box 68, Cambridge, Massachusetts 02139. It is advisable to order a number of extra light bulbs with the apparatus; the life of these bulbs is unpredictable. Note also that much shorter capillaries are required for this apparatus than for most high-grade oil baths. Depending upon the size of capillary stocked, it may be possible to cut the capillaries in half and use each half to make a melting point tube. The capillary tube should project out of the well just far enough to facilitate removal of used capillaries.

Name _____ *Section* _____ *Date* _____

Melting Points

Melting Points of Pure Compounds

Compound name	mp °C (Experimental)	mp °C (Literature)
(a)		
(b)		

Mixture Melting Points

Melting point of a mixture of (a) and (b) _____ °C

Identification of an Unknown

Number of unknown _____

mp of unknown _____ °C

mp of unknown when mixed with _____ _____ °C

(name of compound)

mp of unknown when mixed with _____ _____ °C

(name of compound)

mp of unknown when mixed with _____ _____ °C

(name of compound)

Unknown compound is _____

1. Give two reasons why a sample should be finely powdered and tightly packed when a capillary melting point determination is to be made.

2. What effect would the following nonorganic substances have upon the melting point of an organic compound: (a) sand, (b) water, and (c) salt?

3. What effect would using too large a sample have upon the melting point of a compound?

4. What effect would too rapid heating have upon the apparent melting point of a compound?

5. Some substances, particularly high-melting compounds, give broad, indefinite melting points when the determination is carried out in the usual manner but give sharp, well-defined melting points when the sample is not placed in the bath (or block) until the temperature is about 20° below the melting point of the substance. Suggest an explanation for this behavior.

6. What properties other than those mentioned in the introduction are often valuable as aids in the identification of an organic compound?

7. Name five different natural products of organic origin that lack sharp melting points but soften over a wide temperature range. Are these pure organic compounds or mixtures?

8. One reason why you were instructed not to attempt to redetermine a melting point on a sample previously melted is that the once-melted sample may give either a substantially lower or higher melting point. What sorts of physical or chemical changes could be taking place during melting that would explain this behavior?

9. Phenol, C_6H_6O, is a colorless, crystalline solid with a melting point of 40–41°C, yet the phenol found on a stockroom shelf often appears in the form of a pink slurry rather than as a solid. Offer an explanation for this.

10. How would you determine the melting point of cyclohexane (6.5°) and 1,3-dibromobenzene ($-7°$)?

11. A German chemist refers to glacial acetic acid as *eisessig* (ice vinegar). Why?

12. A very small sample of a low-melting material disappeared from the capillary before the temperature ever reached its reported melting point. Explain.

Recrystallization; Filtration

Organic compounds usually are more soluble in hot solvents than in cold. An impure solid organic compound, when dissolved in the proper amount of an appropriate solvent at an elevated temperature, will reprecipitate when the solution is cooled. If the hot solution is filtered before being allowed to cool, dirt, lint, or other insoluble impurities will be removed, and the crystals that deposit in the cooled solution usually will be more nearly pure than the starting material. The crystals may be removed from the filtrate (mother liquor) by filtration. The soluble impurities and a small amount of the desired substance will stay in solution. This process is called **recrystallization** and is one frequently used for the purification of solid organic compounds. The success of the process depends on the fact that soluble impurities usually are present in smaller amounts than the desired compound so that the cooled solution, although saturated with respect to the desired product, may not be saturated with the impurity. This being the case, the latter will not precipitate from the solution. Sometimes a solution does become saturated with respect to an impurity during the cooling process. In this event, the impurity deposits along with the desired product and the recrystallization will have to be repeated, perhaps several times. In each recrystallization a small amount of the desired compound remains behind in the mother liquor. Such losses are unavoidable if a pure product is to be obtained.

A good solvent for recrystallization has the following properties:

1. It dissolves a reasonable amount of the organic compound at high temperatures, usually the boiling point of the solvent, and very little at low temperatures.

2. It dissolves impurities readily at low temperatures or does not dissolve them at all.

3. It does not react with the substance being purified.

4. It is readily removed from the purified product.

5. It is neither flammable nor toxic.

The fourth requirement means that the solvent must have a relatively low boiling point so that it will evaporate readily. Unfortunately, the last requirement is the hardest to fulfill, for many of the best solvents are either flammable or toxic. Often a solvent with all of these properties cannot be found. In such instances, select the solvent that most nearly approaches the ideal.

The process of recrystallization often may be aided, especially when the impurities are colored, by selectively adsorbing contaminants on activated charcoal. A small amount of charcoal is added to the hot solution just before the filtration step. However, charcoal will adsorb not only the impurities but also a certain amount of the desired product. It is advisable, therefore, to use a minimum amount of charcoal. A small micro-spatula measure usually will be ample for the experiments in this manual.

Inasmuch as many organic compounds are quite bulky and filter slowly, gravity filtration is seldom used in organic chemistry for the collection of precipitates. Suction filtration employing specially designed funnels usually is used to effect rapid filtration and drying of precipitates. For large amounts of material the Büchner funnel (Fig. 2.1) is used. Small amounts of material may be filtered more conveniently by means of a Hirsch funnel (Fig. 2.2).

Figure 2.1   Büchner funnel and filter flask for suction filtration.

Figure 2.2   Hirsch funnel and filter flask for suction filtration.

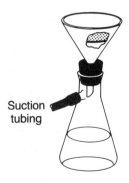

In the organic chemistry laboratory the usual source of vacuum for suction filtration and other techniques requiring moderately low pressures is the water aspirator. The aspirator is a simple, effective device if used properly; however, when many students are using aspirators at the same time, fluctuations in water pressure may result. Sudden drops in water pressure may cause water to be drawn into the vessel being evacuated. Although most aspirators are equipped with check valves to seal the system in the event of such a pressure drop, these valves often fail to function. Therefore, it is advisable to insert a trap (Fig. 2.3) in the line between the aspirator and the filter flask (or other evacuated vessel). This trap is constructed from a 250- or 500-mL suction flask, a stopcock, a length of glass tubing with a right-angle bend, and a two-hole stopper. In the interest of economy you may omit the stopcock; however, it provides easy control over the amount of suction or vacuum in the system and facilitates release of the vacuum at the end of an experiment. If you do not use the stopcock, you should replace it with a short length of glass tubing attached to a short length of rubber tubing. A small clamp on the rubber tubing can then be used as a pinch valve to emulate the functions of the stopcock. Do not replace the suction flask with an ordinary Erlenmeyer flask. Although small Erlenmeyer flasks up to about 50 mL in capacity may be safely evacuated, larger sizes have been known to implode when evacuated. Some current Erlenmeyer flasks are more sturdily built and might be able to stand the vacuum of a water aspirator, but safe practice requires the use of a suction flask. If enough suction flasks are not available, replace the suction flask with a stout cylindrical bottle fitted with a three-hole rubber stopper bearing an inlet tube, an outlet tube, and a stopcock or pinch valve. Note that the inlet tube (the glass tube attached to the rubber hose leading to the aspirator) should reach nearly to the bottom of the trap. All of the flexible tubing between the aspirator and the apparatus being evacuated should be heavy-wall tubing designed for moderate vacuum use. Usually heavy-wall rubber tubing is the most satisfactory.

Always operate the aspirator with the maximum flow of water, using the stopcock or pinch valve to control the amount of suction or vacuum. If several students

Figure 2.3   Trap for water aspirator vacuum systems.

are using aspirators on the same water line, do not leave the apparatus unattended for long periods of time because of the possibility of pressure fluctuations as other students turn their aspirators on and off.

PROCEDURE Frequently it is necessary to heat organic compounds in a solvent for long periods of time without boiling away the solvent. This is accomplished by condensing the hot solvent vapor in a condenser and returning it as liquid to the boiling flask. A continuous boiling and back-flow of condensate in this manner is called **refluxing.**

Figure 2.4 shows an apparatus with a condenser set for refluxing and using a burner as the heat source. If your instructor tells you to use a heating mantle or oil bath as a heat source, refer to the Introduction and Figures 0.1, 0.2, and 0.3 for a discussion of the use of these heat sources and descriptions of the apparatus setups. In this experiment water is the solvent; therefore, the use of a burner to heat the mixture should be quite safe. For recrystallization using flammable solvents, use of a heating mantle or an oil bath, if available, would be preferable.

Figure 2.4 Apparatus assembly for refluxing liquids using standard-taper (⑤) glassware.

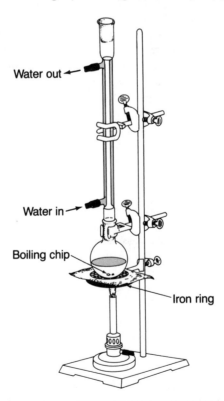

Assemble a reflux apparatus using a 50-mL round-bottomed flask and a small condenser. If you use a burner, support the assembly on a *ceramic fiber-centered wire gauze* on an iron ring about two inches above your burner. Note the positions of the clamps on flask *and* condenser. (Note 1) Make hose connections to the condenser.

Weigh out a 2.0-g sample of impure acetanilide. (Note 2) Remove the reflux condenser temporarily. Save just enough of the impure acetanilide for a melting point determination and add the rest to the 50-mL flask of your semi-macro reflux apparatus. Add one or two boiling chips (Boileezers) to the flask and reassemble the apparatus properly. Add 30 mL of water through the top of the condenser. Attach the lower end of the water jacket of the condenser to a water outlet and the upper end to the sink or drain, using a length of rubber tubing in both cases. A condenser thus is kept filled with cold water simply by allowing it to overflow. (See Fig. 2.4.) Circulate water through the condenser in a slow but steady stream. Bring the water in the small flask to a gentle boil by heating with the burner or other heat source. Adjust the heat source so that the water refluxes in a steady drip from the bottom of the condenser. Continue to heat the mixture until no more solid appears to dissolve. Remove the heat source and *allow the flask to cool a moment*. Then remove the condenser and cautiously add a *very* small amount (match head size on a micro-spatula) of decolorizing charcoal to the hot solution. Replace the condenser and heat source, and boil the mixture five minutes longer.

While the mixture is being heated with charcoal, fit your filter flask with the Hirsch funnel using a tight-fitting rubber stopper. Connect the filter flask to the aspirator with a length of suction tubing. Place a small filter paper disk in the funnel and turn on the water through the aspirator. Preheat the suction apparatus by pouring 50 mL of boiling water through it. Discard the hot water that collects in the filter flask. Turn off the heat source and remove the reflux condenser. Using the flask clamp as a handle, filter the *hot* acetanilide solution without delay. Transfer the filtrate (which probably will already contain some crystals) to a 50-mL beaker and cool by placing the beaker in a pan of ice. Discard the filter disk and the collected impurities, rinse the funnel clean and refit with a new paper filter.

When the filtrate is cold, collect the crystals by suction filtration. Moisten the crystals in the funnel with a little *ice-cold* water and dry them as much as possible with suction. Transfer the crystals to a clean piece of filter paper supported on a watch glass. Cover them with a piece of white paper and store in your desk for drying until the next laboratory period.

Weigh your dried product and determine its melting point as well as that of the crude material. Calculate the percent recovery of pure material. Turn in your purified acetanilide with your written report.

Note 1   Because of the small scale of current elementary organic chemistry equipment a variety of different types of clamps may be used. Two of the more common types are illustrated here, both of which are two-part clamps with a separate two-jawed clamp for attachment to the ring-stand. The upper clamp in Figure 2.4 is a three-jawed condenser clamp, which is very versatile. The lower clamp is a buret clamp.

In many elementary laboratories only buret clamps are provided and, properly used, these are adequate for small-scale apparatus. The clamps should be placed as indicated in the diagrams and adjusted such that the condenser is as vertical as possible and parallel to the rod of the ring-stand. The iron ring should be larger in diameter than the flask above it. Since iron rings are usually not adjustable in the horizontal direction, the buret and condenser clamps may have to be adjusted to center the apparatus over the iron ring.

Note 2 Although it is not necessary to weigh most of the chemicals used in this manual to a high degree of accuracy, it is desirable to be reasonably accurate. For most purposes accuracy of about 1 part per 100 is sufficient. Thus, in this experiment, in order to be able to calculate a reasonable percentage of recovery, your weighing should be accurate to roughly 0.01 g if your balance is capable of such accuracy. If not, try to weigh to a minimum of 0.1 g accuracy.

Crystallization

Amount of pure acetanilide recovered _____ g

Percentage of recovery _____ %

Melting points (°C)		
(impure)	(recrystallized)	(literature mp)

1. A rule early organic chemists frequently followed was stated in scholarly language as *simila similibus solvunter* or "like dissolves like." Restated, this means that a polar solvent will dissolve a polar substance, and a nonpolar solvent will dissolve a nonpolar substance. With this as a guide, what solvent would you use to remove a grease spot from a woolen skirt or from a pair of flannel trousers? What would you use to remove pancake syrup from these garments?

2. Cite several reasons why suction filtration is to be preferred to gravity filtration.

3. Why is a minimum amount of *cold* solvent used for washing precipitates already collected on a filter?

4. A compound is recrystallized from methanol, dried, and examined for purity by determination of its melting point. In a capillary the compound appears to melt at 115° with vigorous evolution of a gas; then the compound solidifies in the hot bath and does not melt again until 165°, at which temperature it again appears to melt sharply. Suggest an explanation for these observations.

5. Why is a mixture of two solvents sometimes necessary for the recrystallization of an organic compound?

6. The formation of crystals from a supersaturated solution, if not spontaneous, may be initiated by simply scratching the inner wall of the vessel containing the solution. Suggest a reason for this.

7. Frequently, a student will attempt to recrystallize a sample from a hot solution only to find that oil globules have formed rather than a homogeneous solution. What has been done wrong and what corrective measures should be taken?

8. Crystals often are separated from the mother liquor by gravity filtration through a filter paper that has been repeatedly folded or "fluted." What is the purpose of using a fluted filter paper?

9. Why should a filtration as described in Question 8 not be used for filtering a hot solution?

10. A crystalline compound is soluble in solvent A and in water. Solvent A is miscible with solvent B, but the material in solution is not soluble in B. The compound is now in solution in A. Suggest a plan for recovering the dissolved material.

11. A substance is recrystallized by dissolving it in hot ethyl alcohol, treating the solution with charcoal, and cooling the filtered solution in an ice bath for an hour. The recovery from this treatment is only 60%. Suggest *at least three* reasons that might explain the low recovery.

Distillation

The space above the surface of a liquid always contains some of the substance in the vapor (gaseous) state. If the container is an open one such as a beaker, the molecules of the vapor escape to the atmosphere to be replaced by other molecules escaping from the liquid. When this process takes place below the boiling point of the liquid it is called **evaporation.** If the container is a closed system such as a corked bottle, an equilibrium is reached in which the air space above the liquid is saturated with molecules of the vapor. The pressure exerted by the vapor in equilibrium with the liquid is called the **vapor pressure** of the liquid. The vapor pressure of a liquid is constant at a given temperature and is not affected by the total pressure. It is a measure of the tendency of the liquid to pass into the vapor state. It increases as the temperature is raised and decreases as the temperature is lowered. The **boiling point** (bp) of a liquid is the temperature at which the vapor pressure becomes equal to atmospheric (total) pressure. Thus, as the atmospheric or total pressure is raised, the boiling point also is raised; as atmospheric or total pressure is lowered, the boiling point is lowered. If the temperature of a liquid is raised until its vapor pressure becomes equal to that of the atmosphere, that is, to the boiling point, and maintained at that temperature, the liquid will pass entirely into the vapor state. This process is known as **distillation.** If the vapor of the liquid is conducted to another container that is cooled below the boiling point, it will **condense** (return to the liquid phase). When chemists speak of distillation they generally mean the combined process of distillation followed by condensation.

Pure organic compounds distill over a very narrow range of temperatures that the organic chemist calls the boiling point. Any liquid with a wide-boiling or distilling range is impure, although a liquid with a very narrow, constant, boiling point range is not necessarily a pure one. The reason for this is that various compounds are affected in different ways by impurities. Some show no change in boiling point, others display elevated boiling points, and still others show depressed boiling points. For this reason the boiling point of a substance, while valuable, is not as good a criterion of purity as is the melting point.

Volatile liquids can be separated from nonvolatile substances by distillation, and, frequently, mixtures of volatile liquids can be separated into the component parts by **fractional distillation.** Occasionally two or more liquids form a constant-

53

boiling mixture or **azeotrope,** which boils at a constant temperature and is made up of a fixed composition of the components. For example, pure ethyl alcohol boils at 78.4° and pure water at 100°, but a mixture of 95% ethyl alcohol and 5% water boils at 78.1°, and this mixture cannot be separated by distillation into pure water and pure ethyl alcohol. Other mixtures having different percentage compositions of ethyl alcohol and water can be separated by distillation into one of the components and the azeotrope.

During distillation a particular portion of the liquid may become momentarily heated above the boiling point, a large amount of vapor may form suddenly, and the contents of the flask may "bump," possibly carrying liquid up through the side-arm of the distilling flask and causing the apparatus to be jolted severely. You can prevent bumping to a large extent by adding a few pieces of porous clay plate or **Boileezers**.[1] These mineral substances contain considerable air in their porous structures, which act as a source of tiny bubbles, first of air, later of the volatile components. The vapors of the liquid form around these bubbles with little local superheating.

The simple distillation apparatus setups described in this manual are used widely in organic chemistry and other laboratories, from the most elementary to the most sophisticated. Because distillation is one of the major techniques used for the isolation and purification of organic compounds, it is important for you to learn to use this technique in a safe and efficient manner. Good technique requires close attention to the progress of any distillation at all times. Before beginning the experiment, review the discussion of Methods of Heating in the Introduction (p. 24), particularly the discussion of the type of heat source that you will use in this experiment. Since in any distillation you will not be heating a constant volume of liquid but a volume that is steadily decreasing, the heating requirements are somewhat different. These differences are discussed here.

Initially, and periodically throughout the distillation, the heat source must be adjusted so that the rate of distillation is neither too fast nor too slow. If the rate is too fast, uncondensed vapors may escape from the condenser, exposing those in the vicinity to the substances being distilled or creating a fire hazard. Observe the liquid condensing in the condenser. If the liquid-vapor interface is much more than halfway down the condenser, turn down the heat or increase the flow of cooling water. The rate of heating may have to be increased during distillation to maintain a reasonable throughput of distillate, especially when mixtures are being distilled. *Toward the end of the distillation, attention is especially important because safety considerations require that the distillation flask (still pot) not be allowed to run dry*. At the end of a distillation, if the rate of heating is still high but little liquid remains to absorb the heat, the walls of the still pot may become very hot, especially if a heating mantle is used. Droplets of condensate that normally fall into the distilling liquid may fall instead on the very hot surface and decompose violently. Also, some organic compounds, especially when stored for a long time, may accumulate traces of heat-sensitive contaminants. Although these contaminants may be relatively in-

[1]Fischer Scientific Company.

nocuous when dissolved in the organic liquid, they may decompose violently when concentrated or when heated in the dry state. Neither of these possibilities is a common occurrence, but both have been known to happen. Therefore, to be safe, toward the end of a distillation *remove or turn down the heat source before the still pot becomes dry*. This is especially important with heating mantles, which retain their heat for some time. Merely turning off a mantle may not suffice; therefore, it is desirable to turn off and then remove the mantle *carefully* while some liquid remains in the still pot.

Although heating mantles are excellent sources of heat for reflux apparatus and the distillation of solvents, they are less satisfactory for general-purpose use in distillation because of the difficulty of regulation and unevenness of heating. Except for the fire hazards associated with open flames, burners are inexpensive, effective sources of heat for many simple organic distillations and have been used for many years in elementary organic chemistry laboratories. Probably the best source of heat for simple distillations is the electrically heated oil bath because the temperature of the bath can be measured with a thermometer inserted into the bath and can be controlled so as not to overheat the still pot. If a hot plate is used to heat the oil bath (rather than an immersion heater), control is somewhat more difficult but can still be accomplished. Since moving a hot oil bath can be hazardous unless a laboratory jack is available, the heat source to an oil bath should be turned off *before* the end of the distillation and the bath not moved unless absolutely necessary.

Caution!   *Stop the distillation by turning off the heat source before the distillation flask runs dry. If you use a heating mantle, remove it after turning it off. If you use an oil bath, remove it if the means to do so safely are provided.*

PROCEDURE

Using Figure 3.1 as a guide, arrange an apparatus for distillation. If heating mantles and/or oil baths are available for your use, you may substitute either for the burner shown in the illustration. If this is done, be sure that you have read the discussion of heating methods given earlier.

The thermometer should be positioned in the upper connecting adapter far enough to allow the tip of the mercury bulb to extend about 10 mm *below* the side-arm opening (see Fig. 3.1). Assemble the apparatus after all glass joints have been lightly lubricated. If a burner is used, support the distilling flask on a ceramic fiber-centered wire gauze. Clamp the flask at the neck. Support the condenser with a second clamp and position it so that it is at the proper height and angle to join the connecting adapter. Make any minor adjustments in clamps and ring-stands necessary to join the condenser to the distilling flask in a strain-free assembly. (Note 1) Attach the take-off adapter to the end of the condenser (a rubber band looped around condenser inlet *a* and vacuum take-off *b* will help keep it in place) and attach a 50-mL receiving flask to this adapter.

Attach the lower end of the water jacket of the condenser to the water line with rubber tubing and the upper end to the drain. Add 1–2 *small* (3–4 mm) pieces of clay plate or Boileezers to the flask and insert the thermometer.

Have your instructor check your apparatus.

Figure 3.1   A simple distillation assembly using standard-taper glassware.

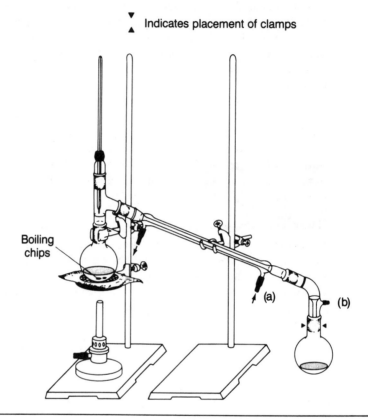

▼
▲   Indicates placement of clamps

Boiling
chips

(a)   (b)

A.   Distillation of a Pure Liquid

Caution!   *Ethanol and 2-propanol are flammable. Make certain that any burners being used are at a safe distance from the distillation receiver, or lead vapors away from the work area as illustrated in Figure 22.1.* (***Instructor: Read Note 1.***)

Pour 25 mL of 2-propanol into the distilling flask using a long-stem funnel so that the liquid will not run down the side-arm. Replace the thermometer, start the water circulating gently through the condenser (a rapid flow is unnecessary), and check all glass-to-glass connections to make certain that they are tight. If you use a burner, adjust it to a low flame and gently heat the flask with the burner until the liquid begins to boil. If you use a heating mantle, adjust the variable transformer until the liquid begins to boil. Remember that it takes some time for the mantle to reach the maximum temperature at any voltage setting, so be careful not to overheat the flask. After you have used a given size of mantle several times, you will learn approximately what settings are required for different temperatures. If you use an oil bath, adjust the heat source until the liquid begins to boil. The temperature of the oil bath may be monitored by means of a thermometer inserted in the bath and supported by a split-ring cork (Fig. 1.1). The temperature of the oil bath will usually have to be 30–50° higher than the boiling point of the substance being distilled; however, the temperature should be kept as low as possible, consistent with a reasonable rate of distillation.

When the liquid begins to drip into the receiver, adjust the heat source so that the distillate drops at a rate of about one drop every two seconds. If you use a mantle as a heat source and the rate of distillation becomes too rapid, it may be necessary to turn off the mantle and lower it temporarily before continuing with a somewhat lower voltage setting. The temperature of an oil bath can be monitored as described above, or the distillation may be observed carefully and the heat source to the bath turned down promptly if the rate of distillation exceeds that recommended.

Record (on the report form or in your notebook) the temperature of the vapor when the first drop of distillate falls into the receiver. Continue to collect the distillate until all but 6–8 mL has distilled. Record the temperature of the vapor again and turn off the heat source. If the heat source is a heating mantle, *lower and remove it cautiously*. If the heat source is an oil bath and you have controlled the temperature of the bath so as to maintain the specified distillation rate and have reduced the temperature to where distillation has stopped and 6–8 mL of liquid still remain in the flask, it should not be necessary to lower the bath until it has cooled. However, *if the bath can be lowered safely with a laboratory jack,* it is probably good practice to do so cautiously.

The two temperatures recorded define the observed boiling point. Return the distillate to the bottle provided on the side shelf. After the residue in the distilling flask has cooled, it may be washed down the drain with copious quantities of water. Discard the used Boileezers in the ordinary trash—not in the sink.

B. Distillation of an Azeotrope

Repeat Procedure A using 30 mL of 25% ethyl alcohol. Remove the small flask used previously as a receiver and collect the distillate in a 25-mL graduate. Record the temperature of the vapor at the first drop and at the end of each 5 mL as the distillation proceeds until you have collected 25 mL of distillate. Record the temperatures on the report form.

Note 1   Recent studies have shown that organic peroxides may form when alcohols, ethers, aldehydes, ketones, alkenes, alkynes, alkyl-substituted arenes, ureas, amides, vinyl monomers, and perhaps other compounds are stored for long periods of time.

These peroxides can decompose violently when heated in distillations, boiling point determinations, and so forth. Therefore, the organic chemicals in each experiment in this manual must be shown by the instructor to be free of peroxide-like materials before use in *any* experiment. Although it may be impractical for students to perform the tests for peroxide content, it is good training to show them how to do so. The instructor may use one of the following tests for this purpose:

(a) To 1 mL of a 20% by weight aqueous solution of potassium iodide, add 1 mL of sample in a small test tube. After vigorous mixing and shaking, a change of color in the aqueous layer from colorless to yellow indicates a low level of peroxide in the sample. A brown-red color indicates high peroxide content. If no color change is observed after several minutes, it is probable that little or no peroxide is present. This test has the disadvantage that the yellow color may be slow to develop, but it is relatively unambiguous.

(b) To a solution of 100 mg of potassium iodide in 1 mL of glacial acetic acid, add 1 mL of sample in a small test tube. A yellow color indicates a low concentration of peroxide, a brown color a high concentration of peroxide. A blank determination *must* be run, since this procedure's disadvantage is that a yellow color may develop even in the absence of peroxide; however, the test is quite rapid.

Alternatively, the instructor may use ether peroxide test strips (Aldrich Chemical Company, catalog number Z10,168-0), following the directions on the package. The test strips should be fairly fresh and should be stored under refrigeration.

Chemicals that give a positive test for peroxide may be purified by passing the material through a chromatographic column packed with activated, neutral alumina. If the amount of peroxide is small (less than 10 ppm by peroxide test paper), 20 g of alumina per 100 mL of liquid may suffice. For high peroxide content more alumina will be required, but it is difficult to estimate exactly how much. In every instance the eluant must be retested for peroxide content, and, if positive, passed through more alumina. Some peroxides are destroyed by the alumina; however, safety considerations require that all traces of peroxides be destroyed by passing a 10–20% solution of ferrous sulfate or of potassium iodide through the alumina before it is discarded.

Note 2   If a three-jawed clamp is used to support the condenser, the two-jawed portion of the clamp should support the underside of the condenser. Adjust this portion of the clamp first (with upper jaw fully open). Then adjust the upper jaw. In disassembling the apparatus, open the upper jaw first, as that will leave the condenser still supported by the lower jaw.

Name   Section   Date

Distillation

Simple Distillation of a Pure Liquid

Volume collected	Temperature, °C
First drop	
20 mL	
Observed boiling point	
Boiling point (literature)	

Distillation of an Azeotrope

Volume collected	Temperature, °C
First drop	
5 mL	
10 mL	
15 mL	
20 mL	
25 mL	

Note   Questions and exercises for this experiment are combined with those for Experiment 4.

Fractional Distillation

A mixture of miscible liquids—that is, each totally soluble in the other but boiling at different temperatures—cannot be separated satisfactorily by simple distillation. The first distillate from such a mixture (called the forerun), although containing some of the higher-boiling material, will always be richer in the lower-boiling component. If the initial distillate is redistilled, the first vapor and condensate again will be richer in the lower-boiling component. The process, if repeated a great many times, will result in a fairly clean separation of the mixture. Obviously, this would be a laborious and time-consuming process. A **fractionating column** is an ingenious device that eliminates the necessity for this multiple manipulation. The fractionating column (Fig. 4.1) is a vertical column packed with some inert material such as glass beads, glass helices, or clay chips to provide a large surface upon which vapor can condense. As the hot vapor rises through this packing it condenses in the cooler part of the column. The condensate flows downward until it reaches a portion of the column sufficiently hot to reconvert it to vapor. Each time the condensate vaporizes, the vapor is again richer in the lower-boiling component than the preceding portion, and the residual liquid in the still pot becomes richer in the higher-boiling component. This process, repeated a great number of times in the packed columns, finally produces vapor of the lower-boiling component that passes as pure compound through the side-arm of the column and into the condenser. The lowest-boiling component will continue to pass over at its boiling point until it is almost completely separated from the mixture. The temperature of the liquid mixture remaining in the flask then will rise to the boiling point of the next lowest-boiling component, and so on until a separation of the liquid mixture is made. This process is called **fractional distillation.** The temperature-composition diagram in Figure 4.2 shows graphically what the fractionating column accomplishes. A mixture of acetone and water in a ratio of approximately 1 : 3 (composition C_1 in diagram) will produce a vapor whose composition V_1 condenses to liquid of composition C_2.[1] The liquid of composition C_2

---

[1] The vapor above a two-component mixture generally will be richer in the lower-boiling component than the liquid. Thus, *liquid* of composition C_1 will have above it *vapor* of the composition represented by the vertical line drawn through V_1 and C_2 to the composition axis and will boil at the temperature indicated by the horizontal line drawn through C_1 and V_1 to the temperature axis.

Figure 4.1 A fractional distillation assembly using standard-taper glassware.

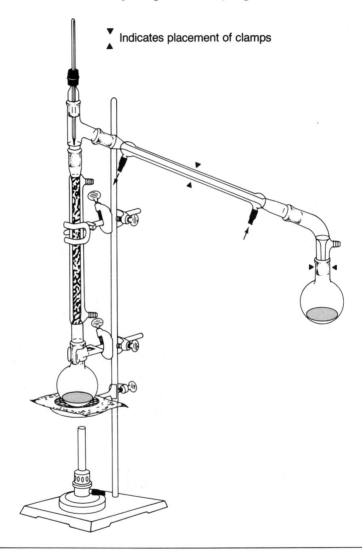

▼
▲ Indicates placement of clamps

produces a vapor of composition V_2. Vapor V_2 condenses to liquid of composition C_3, etc., until finally at point Y on the diagram pure acetone distills.

In general, the smaller the difference between the boiling points of the two liquids in a mixture, the closer together and more nearly horizontal the liquid and vapor curves will be. Thus, there will be a larger number of steps in the schematic representation of the fractionation process, as shown in Figure 4.2b. Obviously a more *efficient* fractionating column would be required to separate two liquids having a small difference in boiling point than that required to separate two liquids having

Figure 4.2   Composition-boiling point diagram for (a) mixtures of acetone and water and (b) a hypothetical mixture of liquids X and Y differing only moderately in boiling point.

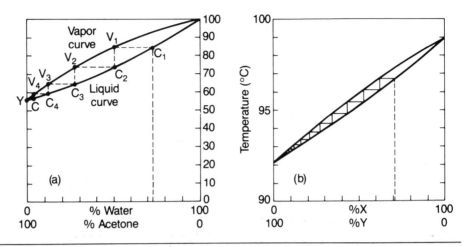

a large difference. The efficiency of any given column can be indicated by specifying the number of *theoretical plates* the column possesses, where a theoretical plate may be regarded as that portion of a column's length required to accomplish one step in the fractionation process. For example, a simple distillation setup (Fig. 3.1) corresponds to roughly one theoretical plate because it can bring about only one step in the fractionation process. Laboratory columns can be constructed having as many as one hundred or more theoretical plates. Although columns having a very large number of theoretical plates are very efficient and can separate liquids boiling only a few degrees apart, the operation of such columns is time-consuming and requires careful attention to operating variables. Simple fractionating columns such as that used in this experiment have at best only a few theoretical plates and are not very efficient, but they suffice to illustrate the principles of fractional distillation. Columns not much more complex than that shown in Figure 4.1 are in routine use in organic chemistry laboratories.

PROCEDURE   ***Caution!*** *If you have not done Experiment 3 or do not remember the discussion concerning the precautions to be taken during distillation, read Experiment 3 before beginning this one. In particular, remember to stop the distillation while there is still a few milliliters of liquid left in the still pot. (**Instructor: Read Note 1, p. 57.**)*

A. Distillation of a Liquid Pair (Without Fractionating Column)

Set up a distillation apparatus as shown in Figure 3.1 using a 50-mL round-bottomed flask as illustrated. Introduce 30 mL of a 50-50 mixture of acetone and water into the distilling flask. Replace the flask normally used as a receiver with a 10- to 25-mL graduated cylinder to facilitate recording of the volume of distillate collected. Add several Boileezers, arrange the thermometer to bring the mercury bulb to the correct position below the side-arm of the adapter, and bring the mixture to a gentle boil. Note and record the temperature at which the first drop of distillate appears. If necessary, adjust the heat during the distillation so that the distillate drips slowly and steadily into the graduated cylinder. Record the temperature every 2 mL as the distillation proceeds until the temperature reaches 99–100° or until all but 6–8 mL of the liquid in the distillation flask has been distilled. Record the volume of the residual liquid in the distilling flask. On the graph provided on the report forms make a plot of temperature vs. volume collected.

B. Distillation of a Liquid Pair (With Fractionating Column)

The apparatus employed for fractional distillation is a modification of that used for simple distillation. Set up a fractionation assembly such as that illustrated in Figure 4.1 using a 50-mL round-bottomed flask, a fractionating column packed with glass beads or some other inert material, and a thermometer. (Note 1) Do not permit the thermometer bulb to come into contact with the packing material of the column, and make certain the bulb is in the correct position below the side-arm opening of the column. For this experiment replace the flask normally used as a receiver with a 10- to 25-mL graduated cylinder as in Procedure A.

Introduce a fresh 30-mL mixture of equal volumes of acetone and water into the flask as in procedure A. Add a Boileezer or two, assemble the apparatus, and bring the mixture to a gentle boil. Note and record the temperature at which the first drop of distillate appears. Record the temperature every 2 mL as the distillation proceeds. As necessary, adjust the heat so that the distillate drips slowly and steadily into the graduated cylinder. Continue distilling until all but 6–8 mL of liquid in the flask has been distilled.

On the graph provided on the report sheet make a plot of temperature vs. volume collected.

Note 1 As in Part A, the heating source may be a Bunsen burner, heating mantle, or electrically heated oil bath.

Name   Section   Date

Distillation of a Binary Mixture

Fraction	First drop	1	2	3	4	5	6	7
Volume (mL)	0	2	4	6	8	10	12	14
Temperature (°C)								

Fraction	8	9	10	11	12	13	Residue
Volume (mL)	16	18	20	22	24	26	
Temperature (°C)							

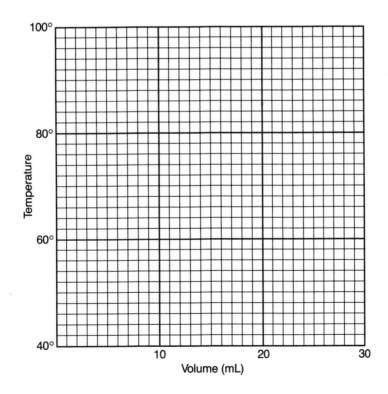

Fractional Distillation of a Binary Mixture

Fraction	First drop	1	2	3	4	5	6	7
Volume (mL)	0	2	4	6	8	10	12	14
Temperature (°C)								

Fraction	8	9	10	11	12	13	Residue
Volume (mL)	16	18	20	22	24	26	
Temperature (°C)							

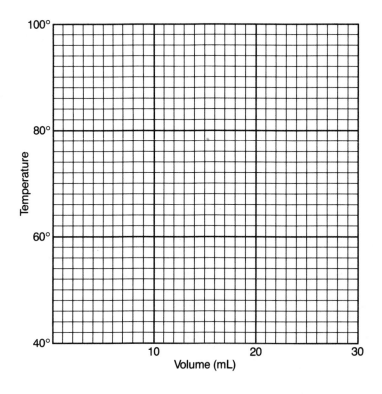

1. If a distillation is started without a Boileezer in the distillation flask, the boiling chip should *never* be added to the hot liquid. Explain.

2. In a beginning organic chemistry laboratory in Denver, Colorado (elevation 5,280 ft), two students distill a pure organic liquid, each student taking his sample from the same bottle. One student reports a boiling point 10° below that recorded in the chemistry handbook; the other reports a boiling point 10° above that in the handbook. One student has reported a proper boiling point; the other has not only reported an improper boiling point but also has probably made a common experimental error. Which student has reported the right boiling point? Why? Which student has reported the wrong boiling point? What was his experimental error?

3. The efficiency of any column packing material can be expressed in terms of its H.E.T.P. (height of an equivalent theoretical plate) rating, where H.E.T.P. can be defined as that length of column packing corresponding to one theoretical plate. Column packings are known with H.E.T.P.'s of less than 1 cm; simple packings such as that employed in Experiment 4 may have H.E.T.P.'s of 10 cm or more. Assume that your packing has an H.E.T.P. of 4 cm and calculate the total number of theoretical plates in your fractional distillation assembly (the complete equipment setup as in Fig. 4.1). To do this, measure the length of the packed section to the nearest centimeter. Some students may turn in answers that will be one theoretical plate low. What is the probable source of their error?

4. It is difficult, if not impossible, to specify exactly how efficient a column will be required to effect the separation of a given pair of liquids, for the answer depends not only on what we mean by "separation" ("pure," 95% pure, or what?) but also on a number of experimental variables. However, we can estimate (in a very approximate way) the number of theoretical plates required by use of a number of empirical relationships that have been devised for this purpose. One simple relationship that has given useful results is the following:

$$n = \frac{(T_1 + T_2)}{3(T_2 - T_1)}$$

where n = the number of theoretical plates, T_2 = the boiling point of the higher-boiling substance (in K), and T_1 = the boiling point of the lower-boiling substance (in K). This equation is intended for use under ideal conditions of "total reflux" (that is, the column at equilibrium but no distillate being collected so that all of the condensed liquid flows back down the column), and a purity of distillate of 95% or better, but it can be used to estimate the minimum number of theoretical plates required for less ideal conditions. Using this equation, estimate how many theoretical plates would be required to separate acetone and water. (Be sure to convert boiling points in °C to K by adding 273.1°.)

5. Pure ethanol boils at 78.4° (760 mm), water at 100.0° (760 mm), and the azeotrope (which contains 95.5% ethanol and 4.5% water) at 78.1° (760 mm). Solutions of ethanol and water containing over 95.5% ethanol can be separated

by fractional distillation into pure ethanol and the azeotrope. Solutions containing less than 95.5% ethanol can be separated by fractional distillation into water and the azeotrope. The azeotrope cannot be separated into its components by fractional distillation. On the basis of this information, how would you expect the vapor-liquid composition curves for ethanol-water to differ from those shown in Figure 4.2? On the graph paper provided, draw a reasonable set of vapor-liquid curves for this system. You may exaggerate the difference in boiling point between pure ethanol and the azeotrope to simplify the drawing.

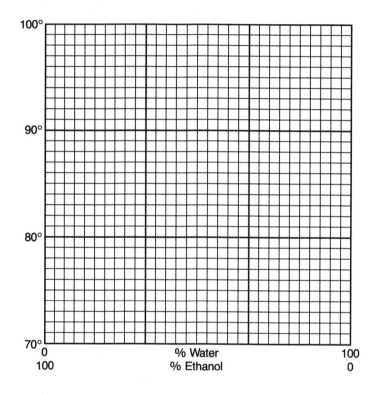

Extraction

Like recrystallization and distillation, extraction is a separation technique frequently employed in the laboratory to isolate one or more components from a mixture. Unlike recrystallization and distillation, it rarely yields a pure product; thus, the former techniques may be required to purify a product isolated in the crude state by extraction. In the technical sense extraction is based on the principle of the equilibrium distribution of a substance (solute) between two immiscible phases, one of which is usually a solvent. The solvent need not be a pure liquid but may be a mixture of several solvents or a solution of some chemical reagent that will react with one or more components of the mixture being extracted to form a new substance soluble in the solution. The material being extracted may be a liquid, a semi-solid, a solid, or a mixture of these. In fact, extraction is a very general, highly versatile technique that is of great value not only in the laboratory but in everyday life. It is probably one of the oldest chemical techniques used by mankind.

As we stated above, the substance being extracted may be a solid or semi-solid. Extractions of this type will not be illustrated in this manual, but they are probably already a part of your own experience. The brewing of tea from tea leaves (or the venerable tea bag that combines extraction and filtration) and of coffee from the ground bean are excellent examples of the extraction of a solid mixture with a hot solvent (water). Other examples include the preparation of vanilla extract from the extraction of the vanilla bean, of gin by the extraction of juniper berries, and of shellac by the extraction of partially purified lac resins obtained from the scale insect, *Coccus lacca*. In each of these examples ethanol is the extracting solvent. Throughout the development of human technology, extraction has been a vitally important process.

In the organic laboratory one of the more important applications of the extraction process has been its use to remove an organic compound from a solution when distillation is not feasible or advantageous. Extraction is accomplished by shaking the solution in a separatory funnel with a second solvent that is immiscible with the one in which the compound is dissolved, but one that dissolves the compound more readily. Two liquid layers thus are formed, and the layer that has most of the desired product in it can be separated from the other. Sometimes not all of the product is extracted in a single operation and the process must be repeated once or twice more to assure a clean separation. It has been found that when two immiscible solvents are shaken together, the solute distributes itself between them in a ratio roughly proportional to its solubility in each. The ratio of the concentration of the solute in each solvent at equilibrium is a constant called the **distribution ratio** or **distribution coefficient.** For example, at 20° only 0.24 g of azelaic acid will dissolve in 100 mL of water, but 2.70 g of the same acid will dissolve in 100 mL of ether. When shaken with a mixture of equal volumes of water and ether, azelaic acid will distribute between the water and ether layers so that the concentration of azelaic acid in the ether layer will be slightly more than ten times that in the aqueous layer. The exact value of the distribution ratio, K_d in this case, would be the same as the ratio of the solubilities only if each solvent were *completely* immiscible. This seldom is the case. For practical purposes, however, an *approximate* value of the distribution ratio may be calculated from the following equation,

$$K_d = \frac{C_e}{C_w} = \frac{W_e/100 \text{ mL}}{W_w/100 \text{ mL}} = \frac{2.7}{0.24} = 11.25$$

where C_e and C_w are the concentrations in the ether and water layers respectively, and W_e and W_w are the weights in grams of material dissolved in each respective layer. One can easily calculate the amount of material extracted by a given volume of solvent and how much solute remains in the aqueous layer if the numerical value of the distribution ratio is known. For example, if 0.12 g of azelaic acid in 100 mL of aqueous solution were extracted with 100 mL of ether, the weight (W) of acid extracted by the ether may be calculated as follows:

$$\frac{\dfrac{W}{100}}{\dfrac{(0.12 - W)}{100}} = 11.25; \; W = 0.1102 \text{ g } (92\%)$$

To determine the amount of material that might have been extracted by using only 50 mL of ether, a similar calculation is made.

$$\frac{\dfrac{W'}{50}}{\dfrac{(0.12 - W')}{100}} = 11.25; \; W' = 0.102 \text{ g } (83\%)$$

A second 50-mL ether extraction of the residual aqueous solution would remove another proportionate amount of solute according to the distribution ratio.

$$\frac{\dfrac{W''}{50}}{\dfrac{(0.12 - 0.102 - W'')}{100}} = 11.25; \; W'' = 0.0153 \text{ g } (12.8\%)$$

From these calculations you can see that a multiple extraction, using smaller volumes of extracting solvent, more efficiently removes a solute from solution than does one extraction with a much larger volume of solvent.

A second important application of the extraction process in the organic chemistry laboratory involves the use of what is often called a **reaction solvent,** where a reaction solvent can be defined as a *solution* of some reagent (in an appropriate solvent) that reacts selectively with one or more components of a mixture to form a new substance(s) soluble in the solvent. To be effective, the reaction solvent must dissolve only those compounds that react with the reagent. For example, consider a mixture consisting of an organic acid

$$R-C\underset{OH}{\overset{O}{\diagup}}$$

an organic base or amine ($R'-NH_2$), and a neutral organic hydrocarbon ($R''-H$), all of which are water-insoluble. The mixture is dissolved in some convenient solvent that is immiscible with water and has a reasonably low boiling point. If the solution of the mixture is first extracted with a dilute (5–10%) aqueous solution of hydrochloric acid, only the amine will react to form the water-soluble organic ammonium chloride.

$$R'-NH_2 + HCl \longrightarrow R'-NH_3^+ + Cl^-$$

Thus, the amine passes from the organic solvent layer into the aqueous layer in the form of the ammonium salt. If the aqueous layer is separated and neutralized with base (e.g., sodium or potassium hydroxide), the insoluble amine will precipitate.

$$R'-NH_3^+Cl^- + NaOH \longrightarrow R'-NH_2 + NaCl$$

Now, if the organic solvent layer (still containing the acid and the hydrocarbon) is extracted with a dilute aqueous solution of sodium hydroxide, only the acid will react to form the water-soluble sodium salt.

$$R-C\underset{OH}{\overset{O}{\diagup}} + NaOH \longrightarrow R-C\underset{O^-}{\overset{O}{\diagup}} + Na^+ + H_2O$$

Thus, the acid now passes from the organic solvent layer into the aqueous layer in the form of the sodium salt. If the aqueous layer is separated and neutralized with acid (e.g., dilute hydrochloric acid), the insoluble acid will precipitate.

$$R-C\underset{O^-Na^+}{\overset{O}{\diagup}} + HCl \longrightarrow R-C\underset{OH}{\overset{O}{\diagup}} + NaCl$$

At this later stage we have three mixtures: (1) organic amine plus water, (2) organic acid plus water, and (3) a neutral compound in some organic solvent. If either the acid or the amine is solid and highly water-insoluble, it may be recovered by simple suction filtration. However, assume that neither the amine nor the acid can be recovered this simply. Then it will be necessary to extract the water-amine and water-acid mixtures with an appropriate organic solvent to separate these substances from the water. After extractions we have three solutions, each containing one of the three components of the original mixture in an organic solvent. Each component may be recovered by evaporation or, if necessary, fractional distillation of the solvent. Of course, each will probably have to be further purified by recrystallization or distillation.

Sometimes it is helpful in planning a separation scheme based on extraction (alone or with other separation techniques) to prepare a flow chart outlining the various stages in the process. Thus, the example described above could be outlined as follows, assuming that the organic solvent is ether.

$$
\left.\begin{array}{l}
R-CO_2H \\
R'-NH_2 \\
Neutral
\end{array}\right\}
\xrightarrow[\text{water}]{\text{HCl}}
$$

Water layer $\longrightarrow$ $R'-NH_3{}^+Cl^-$ $\xrightarrow{\text{NaOH}}$ $R'-NH_2$

Ether layer $\left\{\begin{array}{l} R-CO_2H \\ Neutral \end{array}\right\}$ $\xrightarrow[\text{water}]{\text{NaOH}}$

Water layer $\longrightarrow$ $R-CO_2{}^-Na^+$ $\xrightarrow{\text{HCl}}$ $R-CO_2H$

Ether layer $\longrightarrow$ Neutral

Ether

From the foregoing discussions some of the desirable properties of an organic extraction solvent become apparent. It must readily dissolve the substance being extracted but must not dissolve to any appreciable extent in the solvent from which the desired substance is being extracted. It should extract neither the impurities nor other substances present in the original mixture. Except in the case of reaction solvents, it should not react with the substance being extracted. Finally, it should be readily separated from the desired solute after extraction. Few solvents will meet all of these criteria, and in some cases no completely satisfactory solvent can be found. Therefore, as in the instance of recrystallization, the chemist will select that solvent that most nearly approaches the ideal.

Some of the solvents commonly used for extracting aqueous solutions or mixtures include diethyl ether, methylene chloride, chloroform, carbon tetrachloride, benzene, *n*-pentane, *n*-hexane, and various mixtures of saturated hydrocarbons from petroleum (petroleum ether, ligroin, etc.). Each of these has a relatively low boiling point so that it may be fairly easily separated from the solute by evaporation or distillation. Methanol and ethanol are not good solvents for extracting aqueous solutions or mixtures because of their solubility in water; however, if an aqueous solution can be saturated with potassium carbonate without affecting the solute, ethanol can be used to extract polar solutes from the solution.

Caution! *Benzene, carbon tetrachloride, and chloroform, and chlorinated hydrocarbons, in general, are cancer suspect agents. Do not use them as solvents in this laboratory without your instructor's specific approval.*

Use of the Separatory Funnel

The procedure in this experiment involves the use of the separatory funnel. It is important that you learn how to use this rather expensive piece of equipment properly, for, in addition to its use for the extraction of organic substances from aqueous solutions, the separatory funnel is often employed for washing organic liquids. It is made of thin glass and is easily broken unless handled carefully. Unfortunately, although most organic chemists are in agreement about most aspects of the technique for *using* the separatory funnel, there is almost no agreement about the best way of *holding* it. In various student manuals you will find descriptions of about as many ways of holding the separatory funnel as for holding a baby, a golf club, or a pair of chopsticks. Even the two authors of this manual use slightly different "grips." Probably for you there is some best method, depending on the size of your hands, the strength in your fingers, your manual dexterity, and the size and shape of the funnel. The following are important rules to observe.

1. Hold the funnel firmly but gently in both hands so that it can be turned from the vertical to horizontal direction and back again easily and can be shaken vigorously while observing factors (2) and (3).

2. Keep the stopper tightly seated with one hand at all times, using the forefinger of that hand, the base of the forefinger, or the palm of the hand.

3. Keep the stopcock tightly seated with the fingers of the other hand in such a way that the fingers can open and close the stopcock quickly to release the pressure that may be built up from solvent vapor or evolved gases such as nitrogen or carbon dioxide.

Obviously, the use of the separatory funnel is a skill and is best learned by practice with an empty funnel while watching an experienced person such as your instructor demonstrate his or her technique. Do not forget to lubricate both the stopcock and the stopper before working with the empty funnel. In Figure 5.1 are

Figure 5.1   Two methods for holding the separatory funnel.

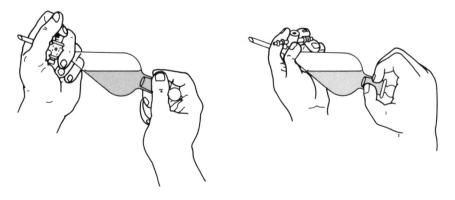

shown two slightly different methods of handling the separatory funnel taught to students in our laboratories. In the first method the stem of the funnel projects between the thumb and first finger of the left hand (for a right-handed person). The stopcock is held in place and operated with the thumb and first finger. The stopper is kept in place by pressure against the base of the first finger of the right hand.

In the second method the stem of the funnel projects between the first and second fingers of the left hand. The stopcock is held in place by pressure from these fingers and is operated by them in conjunction with the thumb. The stopper is held in place by pressure against the middle of the palm of the right hand. In other common variants not shown, the stopper may be held in place by pressure from the tip of the first finger, and the use of the hands may or may not be reversed. Confusing? Not really. Your instructor will help you select a method that works best for you and your equipment. Once you have developed your "grip" with the empty funnel, the following procedure is the one most organic chemists use.

Support the separatory funnel in a plastic coated iron ring or one padded with short sections of rubber tubing split lengthwise and wired to the ring. Lubricate the stopcock of the funnel (unless it is made of Teflon, in which case it will require no lubrication). Close the stopcock and add to the funnel the liquids to be separated. Insert the stopper, which should be greased lightly with stopcock grease, and immediately invert the funnel. Point the barrel away from your face and that of your neighbor. Open the stopcock to release the pressure, which may have accumulated inside the funnel (volatile solvents such as ether develop considerable pressure).

Figure 5.2   Support and use of the separatory funnel.

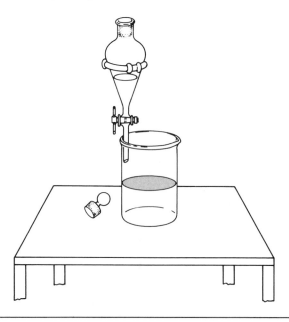

Close the stopcock and, holding the funnel horizontally, shake the funnel two or three times. Invert the funnel and release the pressure as before. Repeat this process until opening the stopcock causes no further pressure release. Close the stopcock and shake the funnel 15–20 times. Replace the funnel in the iron ring and *remove the stopper*. Allow the liquids to stand until the layers have separated sharply. Draw the lower layer into a flask or beaker of proper size (Fig. 5.2). Do not draw the liquid through the stopcock too rapidly. Slow the flow carefully as the boundary between the two layers approaches the stopcock. Stop the flow of liquid completely just as the upper layer enters the hole in the stopcock. Pour the upper layer *through the neck* of the funnel into a second flask. *Never discard either layer until you are absolutely certain which is the proper layer to keep.* Usually one layer will be an aqueous layer or solution, and the other will be an organic liquid. The one of greater density, of course, will be on the bottom.

To check the identity of a layer, should you be in doubt, withdraw a few milliliters of the lower layer into a test tube containing an equal volume of water. If the lower layer in the separatory funnel is water or an aqueous solution it will be homogeneous (only one layer). If the layer being tested is the organic layer, the sample withdrawn will fall to the bottom of your test tube and also form two liquid layers. In either event, return the test mixture to the separatory funnel.

> **Precautions.** *This experiment requires the use of hydrochloric acid, certain organic acids, ether, sodium hydroxide, and aromatic amines. See HAZARD CATEGORIES 1, 2, 3, and 8, page 4.*

PROCEDURE   A. Extraction of Mandelic Acid from Water

In a 250-mL Erlenmeyer flask dissolve 8 g of mandelic acid, $C_8H_8O_3$, in 100 mL of distilled water.

Step 1 Use your graduated cylinder to measure 30 mL of the acid solution and transfer the solution to a 125-mL Erlenmeyer flask. Add 3–4 drops of phenolphthalein and titrate to the end point with a standardized (about 0.3 N) sodium hydroxide solution using one of the buret assemblies provided. Record on your report sheet the number of milliliters of base required to neutralize this volume of acid solution. Calculate the exact weight of mandelic acid dissolved in 30 mL of aqueous solution and also calculate the number of grams of mandelic acid neutralized by each milliliter of the standard base. Discard the neutralized acid solution and rinse your flasks.

Step 2 Measure out a second 30-mL volume of acid solution and transfer it to your separatory funnel. Add 30 mL of ether to the funnel and extract according to the procedure outlined in the introductory section of this experiment. Separate the *lower* aqueous layer into a 125-mL Erlenmeyer flask and add 3–4 drops of indicator. Record the volume of the sodium hydroxide solution in the buret and titrate to the phenolphthalein end point. Again record the number of milliliters of base required

and calculate the number of grams of acid removed by the ether and the number of grams remaining in the aqueous layer. Discard the neutralized acid solution and empty the ether layer into the large bottle marked "Solvent Ether from Extractions."

Step 3 Repeat the procedure from Step 2, but this time extract 30 mL of fresh acid solution with only 15 mL of ether. Separate the aqueous layer into a flask and dispose of the ether layer. Transfer the aqueous layer back into the empty, cleaned, separatory funnel and extract it with a second 15-mL portion of fresh ether. Separate the extracted aqueous layer, add indicator as before and titrate to the end point. Record the volume of standard base required and calculate how much acid remains in the aqueous layer and the total acid removed by the combined ether extracts. Dispose of the ether extracts as directed, clean your separatory funnel, and store your equipment.

B. Separation of a Mixture by Extraction

In this part of the experiment a mixture containing an acidic, a basic, and a neutral compound is to be separated into the individual components, which will be chosen from the substances whose formulas are shown below.

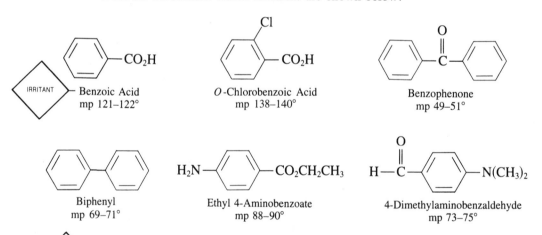

Benzoic Acid
mp 121–122°

IRRITANT

O-Chlorobenzoic Acid
mp 138–140°

Benzophenone
mp 49–51°

Biphenyl
mp 69–71°

Ethyl 4-Aminobenzoate
mp 88–90°

4-Dimethylaminobenzaldehyde
mp 73–75°

In a 125-mL Erlenmeyer flask dissolve 3.0 g of the three-component mixture your instructor provides in 25 mL of ether. Pour the solution into your separatory funnel. Rinse the flask with 5 mL of ether and add the ether to the separatory funnel. Extract the organic base (amine) from the mixture using two 25-mL portions of 5% hydrochloric acid (Note 1) and following the procedure outlined in the introductory section of this experiment. Draw off the *lower* aqueous layer in each extraction into the same 125-mL Erlenmeyer flask. Extract the ether layer with 10 mL of water and add the water to the combined HCl extracts. Label the flask "HCl Extract" and set it aside.

Next, extract the organic acid from the mixture, using two 25-mL portions of 5% aqueous sodium hydroxide (prepare by diluting 25 mL of 10% sodium hydroxide solution with 25 mL of water) and one 10-mL portion of water, again drawing all of the lower aqueous layers into one 125-mL Erlenmeyer flask. Label the flask "NaOH Extract" and set it aside.

Pour the ether layer in the separatory funnel *through the neck* of the funnel into a 50-mL Erlenmeyer flask and add 1 g of anhydrous magnesium sulfate. Allow the ether solution to stand for 10–15 minutes, swirling occasionally. Filter the solution by gravity through a small plug of cotton placed in the bottom of a conical funnel (Note 2) into a 100-mL beaker. Rinse the magnesium sulfate with 1–2 mL of ether, allowing the ether to run into the beaker. Place the beaker in a good fume hood (**No flames!**) and allow the ether to evaporate. The residue should be your neutral compound. Weigh the compound and determine its melting point and, if your instructor directs, determine its mixture melting point with some of the pure substance to establish its identity. (Note 3)

Cool the HCl Extract in the ice bath and make it basic by adding 10% aqueous sodium hydroxide until the solution is distinctly alkaline to litmus. Chill the mixture thoroughly and collect the released organic base (amine) by suction filtration, using the Hirsch funnel. (Note 4) Dry the product in air, weigh it, determine its melting point and, if directed, its mixture melting point with some of the pure substance to establish its identity.

Cool the NaOH Extract in the ice bath and make it acidic by adding concentrated hydrochloric acid dropwise until the solution is acid to Congo Red paper. (Note 5) Chill the mixture thoroughly and collect the released organic acid (Note 4) by suction filtration, using the Hirsch funnel. Dry the product in air, weigh it, determine its melting point and, if directed, its mixture melting point with some of the pure substance to establish its identity.

Note 1   It is advisable to use moderately dilute solutions of acids and bases for extraction purposes; 5% aqueous solutions appear to be a reasonable compromise. Approximately 5% hydrochloric acid may be prepared by adding 6 mL of concentrated hydrochloric acid to 50 mL of water.

Note 2   The cotton plug should be no larger than the tapered end of a pencil and must fit snugly in the bottom of the funnel but should not be compressed so tightly that the liquid does not flow through freely. It may be advisable to wet the cotton with a small amount of ether before filtering the solution. It is also good practice to decant as much of the solution as possible into the funnel before adding the mixture of solution and magnesium sulfate.

Note 3   Dry and weigh all compounds recovered and take their melting points before the end of the laboratory period. If kept uncovered until the next laboratory period before melting points are taken, the organic base may oxidize.

Note 4   If there is any doubt about the toxicity of the vapors of organic acids or bases being recovered from an extraction, perform the filtration step in a good hood to minimize exposure to the vapors of the materials. In this experiment this should not be necessary, although if sufficient hood facilities are available, it is always good practice to do as much chemistry in them as possible. Usually in extractions the most important safety measure to observe is to avoid physical contact with the chemicals or solutions.

Note 5   Congo Red changes color in the pH range of approximately 3 to 5, being red in neutral or alkaline solution and blue in acidic solution. It is particularly useful in the organic laboratory because this range is just below that of aqueous solutions of most carboxylic acids. Therefore, Congo Red paper generally remains red in the presence of carboxylic acids but changes to blue in the presence of mineral acids (such as hydrochloric and sulfuric acids). When a solution is "acidified to Congo Red" or "made acidic to Congo Red," just enough mineral acid is added to cause the indicator paper to turn from red to blue. Be sure that the paper is wet by the aqueous solution and not just by the organic layer. If Congo Red paper is not available, the more expensive but very versatile Hydrion pH papers may be used. Choose either the "Insta-Chek" general purpose paper or a "Microfine" paper in the 2–4 pH range. Occasionally, the pH may have to be brought as low as 1–2 in order to recover the stronger organic acids.

Extraction of Mandelic Acid from Water

Step 1   Volume of base required to neutralize 30 mL of acid. _____ mL

(a) Number of equivalents of acid in 30 mL of solution.

$$\frac{\text{mL of base}}{1{,}000} \times N \text{ (base)} = \underline{\hspace{2cm}} \text{ Eq.}$$

(b) Grams mandelic acid in 30 mL of solution.

$$(1a) \times 152 \text{ g (MW mandelic acid)} = \underline{\hspace{2cm}} \text{ g}$$

(c) Grams acid neutralized by 1 mL of base. _____ g

Step 2   (a) Number mL of base used. _____ mL

(b) Number of grams of acid remaining in aqueous layer after one 30-mL ether extraction

$$(2a) \times (1c) = \underline{\hspace{2cm}} \text{ g}$$

(c) Number of grams acid removed by one 30-mL portion of ether.

$$(1b) - (2b) = \underline{\hspace{2cm}} \text{ g}$$

(d) Percent of mandelic acid extracted.

$$(2c) \times 100 \div (1b) = \underline{\hspace{2cm}} \%$$

Step 3   (a) Number of mL of base used. _____ mL

(b) Number of grams of acid remaining in aqueous layer after two 15-mL ether extractions.

$$(3a) \times (1c) = \underline{\hspace{2cm}} \text{ g}$$

(c) Total acid removed by both ether extracts.

$$(1b) - (3b) = \underline{\hspace{2cm}} \text{ g}$$

(d) Percent of mandelic acid extracted.

$$(3c) \times 100 \div (1b) = \underline{\hspace{2cm}} \%$$

Separation of a Mixture by Extraction

Unknown Sample No. _____

1. Neutral Compound

   Weight of neutral compound: _____ g

   mp of neutral compound: _____ °C

   Mixture mp of compound with _____ : _____ °C

   Mixture mp of compound with _____ : _____ °C

   Identity of neutral compound: _____

2. Acidic Compound

   Weight of acidic compound: _____ g

   mp of acidic compound: _____ °C

   Mixture mp of compound with _____ : _____ °C

   Mixture mp of compound with _____ : _____ °C

   Identity of acidic compound: _____

3. Basic Compound

   Weight of basic compound: _____ g

   mp of basic compound: _____ °C

   Mixture mp of compound with _____ : _____ °C

   Mixture mp of compound with _____ : _____ °C

   Identity of basic compound: _____

1. What are the advantages and disadvantages of using ether as a solvent for the extraction of organic compounds?

2. What volume of an organic solvent must be used to effect a 90% separation in one extraction when only 2.7 g of a certain organic compound dissolves in 100 mL of water? ($K = 15$)

3. What percentage of the organic compound could be recovered if two extractions were made, each time using half the volume calculated in Question 2?

4. An organic compound can be extracted from a water layer by an organic solvent more efficiently if the water layer is saturated with an inorganic salt such as sodium chloride. This effect, called "salting out," increases the partition coefficient in favor of the organic compound. Explain.

5. Carboxylic acids (R—COOH) are relatively strong acids and will dissolve in both 5% aqueous sodium hydroxide and 5% aqueous sodium bicarbonate. Phenols (Ar—OH) are weak acids, and most simple phenols will dissolve in 5% aqueous sodium hydroxide (forming the sodium salts, Ar—O⁻Na⁺) but will not dissolve in 5% aqueous sodium bicarbonate (that is, they do not react with sodium bicarbonate). With this information and that in the introduction to this experiment, work out a flow chart for the separation of a carboxylic acid (R—COOH), an amine (R—NH₂), a phenol (Ar—OH), and a neutral hydrocarbon (R—H). Assume that all four compounds are insoluble in water and soluble in ether.

6. In addition to the properties of a low boiling point and immiscibility with water, why is density important to consider in the selection of an extraction solvent? Would an organic liquid with a density of 1.008 g/mL make a good extraction solvent? Why?

7. In Part B of our experiment, why was a solution of sodium bicarbonate not used as a reaction solvent for the extraction of the acid component of the mixture?

8. You are probably familiar with the standard handbooks of chemistry and physics; however, you should become familiar with some of the other one-volume reference works available, particularly the *Merck Index*, which is of interest to persons in the biological and medical sciences. All of the organic acids and bases listed in Part B of this experiment are listed in the *Merck Index*. Look them up and report any interesting properties that you find for each. Which of these compounds might be used in a sunburn cream to alleviate the pain of sunburn?

Sublimation:
The Isolation of Caffeine from Tea

Caffeine, whose structure is shown, occurs in coffee, tea, and cola drinks. It is a heterocyclic compound belonging to a broad class of naturally occurring, basic, nitrogenous plant products known as **alkaloids.** Although most alkaloids have a profound physiological action when administered to animals, the effect of caffeine upon the nervous system is that of a mild stimulant.

Caffeine

Caffeine is present in tea leaves to the extent of 4–5%. It is soluble in hot water, but soluble in only a few of the common organic solvents. In the present experiment we will extract caffeine from tea leaves with hot water much as is done in the home but in a greater concentration. After a second extraction of the cold aqueous tea solution with an organic solvent, the crude caffeine is concentrated and purified by sublimation.

Sublimation

A number of solid organic compounds will produce high vapor pressures at temperatures well below their melting points. Such solids, when heated, pass directly into a vapor without first passing through the liquid state. The process is reversible and the vapors, when cooled, will recondense directly to the solid. The process of converting a solid directly into its vapor and the condensation of the vapor back to the solid is called **sublimation.** Sublimation is one of the best methods for obtaining an organic compound in a high state of purity or for removing a compound as a sublimate from nonvolatile contaminants.

Figure 6.1 Sublimation apparatus.

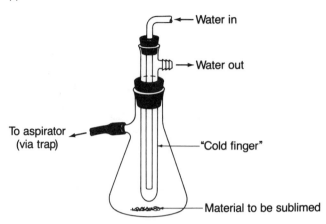

Caffeine has a melting point of 237°, but will sublime at much lower temperatures under the diminished pressure afforded by an efficient water aspirator.

A number of different types of sublimation apparatus are available commercially, but a simple one made from ordinary laboratory ware is illustrated in Figure 6.1. A 150 mm × 18 mm test tube with a side-arm and fitted with a one-hole rubber stopper through which a short length of 6 mm glass tubing has been inserted makes a satisfactory condenser called a "cold finger." The "cold finger" is held in a 125-mL filter flask by means of a snug-fitting rubber stopper that has been bored to proper size. The bottom of the "cold finger" is positioned about 1.0 cm above the bottom of the filter flask. The material to be sublimed is placed in the filter flask and heated to a temperature below its melting point while pressure is reduced by means of a water aspirator.

PROCEDURE

IRRITANT

Into 300 mL of rapidly boiling water contained in a 1-liter beaker slowly immerse 8–10 tea bags (approximately 20–25 g). Allow to stand for 10 minutes, agitating and pressing the tea bags once or twice with a stirring rod. Using a towel for easy handling, carefully decant the tea solution along with the spent tea bags into a 10-cm Büchner funnel that has *not* been fitted with a filter paper. Express as much solution from the bags as you can by using the base of a small Erlenmeyer flask as a press. Allow the filtrate to come to room temperature and then transfer it to a separatory funnel. Add 25 mL of methylene chloride and swirl gently. (Note 1) Allow the two layers to separate and draw off the lower organic layer into a small Erlenmeyer flask and set aside. (Note 2) Make two additional extractions of the aqueous layer, each time using a 25-mL volume of methylene chloride. Combine these extracts with the first. Discard the dark brown aqueous layer, rinse the separatory funnel with water, and return to it the combined methylene chloride extracts. Add 50 mL of saturated sodium chloride solution to the separatory funnel and gently agitate the two layers. The upper aqueous layer should now be amber-colored and the lower layer should be clear. (Note 3) Separate the organic layer into a small flask and add 10 g of anhydrous magnesium sulfate or enough to freely flow along the bottom of the flask and not form lumps. While the methylene chloride solution is drying, assemble a distillation apparatus such as that illustrated in Figure 3.1, but with a stopper in place of the thermometer adapter. Transfer the dried methylene chloride solution to the distilling flask via a long-stemmed funnel fitted with a small plug of cotton, add a Boileezer, and remove by distillation all but 10 mL of the solvent. Use a heating mantle or a hot water bath as your heat source. (Note 4) The boiling point of methylene chloride is 41° and only mild heating is required. Transfer the residual 10 mL of solution from the boiling flask to the filter flask of the sublimation apparatus. Include a trap such as the one illustrated in Figure 2.3, insert the "cold finger," and evaporate the solution to dryness under diminished pressure. Use either a hot plate or a Bunsen burner adjusted to a *very low* flame as your heat source. (Note 4) Allow water to flow through the "cold finger" only after all solvent has been removed. Sublime the caffeine by applying heat to the bottom of the filter flask while at the same time applying a suction with the aspirator. Carefully remove the "cold finger" after sublimation is complete and scrape off the accumulated caffeine along with that which will have condensed on the side of the flask and determine your yield. (Note 5) Take a melting point in a sealed capillary (Note 6) or with a Fisher-Johns apparatus with a sample sandwiched between cover glasses. The melting point of caffeine is reported in the literature as 237°.

Caution! *The melting point of caffeine is too high to take its melting point in an oil-filled melting point bath unless the bath is filled with silicone oil.*

Note 1   Chlorinated hydrocarbons as extraction solvents have an annoying tendency to form emulsions, especially when shaken vigorously. In this extraction hold the separatory funnel in a horizontal position and gently rock it back and forth. Although methylene chloride is classified as a toxic irritant, there is some concern that it may be a mild carcinogen. Your instructor will inform you of any change in classification.

Note 2   There will not be a clean separation of layers. The *complete* lower layer should be separated from the upper dark layer even though the lower layer will contain some emulsion.

Note 3   There is still a tendency toward emulsion formation. With time and occasional gentle swirling, a minimum of emulsified material will be withdrawn with the organic layer.

Note 4   Methods of heating are discussed in the General Instructions beginning on page 24 and in Experiment 3 beginning on page 53.

Note 5   You can make a more accurate determination of yield by weighing the sublimation apparatus before it is charged and again with the sublimate. A yield of approximately 1 g is satisfactory.

Note 6   The instructor can prepare sealed capillaries easily with the aid of an air-gas glass-blowing torch. The melting point capillary is first filled and tamped in the usual manner. Then the tube is sealed about 2 cm above the packed portion, using a small flame and working as quickly as possible. Avoid overheating the sample while sealing the capillary. If this proves difficult, try sealing a little farther up the capillary or wrapping the packed end of the capillary in a scrap of filter paper saturated with ice water.

REPORT
6

Name *Section* *Date*

Theoretical recovery of caffeine ________ g

Actual recovery ________ g

Percentage recovery ________ %

mp ________ °C

1. Consult your text and name the heterocyclic ring system of which caffeine is a derivative.

2. Caffeine may also be named 1,3,7-trimethylxanthine. What is the structure of xanthine? Number the ring system.

3. Name some familiar nonorganic substances that sublime readily.

4. Name some organic compounds whose usefulness in the home depends upon their ability to sublime.

5. A storage chest for woolen blankets measures 18 in × 18 in × 42 in. How many grams of naphthalene flakes, when completely sublimed, would fill this chest with moth-repellent vapor? (1 in = 2.54 cm)

6. What was the reason for the instructions given at the end of the experimental procedure regarding the taking of a melting point?

7. An organic compound has a melting point of 179°. Its vapor pressure at this temperature is 370 mm. Tell how you might purify this compound by sublimation. What would be a convenient source of heat for this phase change?

8. Would you believe the following statement to be true or false? Explain. "Any solid organic compound of pronounced odor may be sublimed."

Chromatography

Chromatography (Gr., *chroma*, color; *graphein*, to write) is a technique chemists frequently employ to separate the components of a mixture. In its original application the method was used for the separation of colored substances, but color need not be a required property for compounds to be separated by chromatography. Colorless compounds may be rendered visible by other means. A number of separation techniques embodying the principles of chromatography have been developed. Four of these, along with one or more experiments for each, will be described.

PART I Column Chromatography

A mixture to be separated by **column chromatography** is introduced as a solution into the top of a vertical column packed with some finely divided, inert material already wet with an adsorbed solvent. Enough fresh solvent is added continuously to the top of the column to carry the mixture down and through the supporting material. Each component of the mixture, in descending through the column packing, partitions itself many times between the adsorbed solvent (called the **stationary phase**) and the moving solvent (called the **mobile phase**). The distribution coefficient with respect to these two phases is different for each compound comprising the mixture, and, therefore, each travels through the supporting material at a different rate. Components of the mixture thus become separated within a short time. If the mixture is one of colored substances, compounds appear in the column packing as distinct, colored bands. Continued addition of solvent to the column finally elutes each compound as a colored solution out the lower end and into a receiver. The recovery of each compound as a pure substance then is possible by simply evaporating the solvent. This type of column chromatography is referred to as **partition chromatography.**

When the column packing is wet with the same solvent as that used to elute the mixture, the rate of descent through the column is dependent upon how strongly each component is adsorbed to the column packing. This type of column chromatography is referred to as **adsorption chromatography.** In either case, differences in solubility behavior or differences in adsorption properties are functions of the molecular structures of the different compounds to be separated.

> **Precautions.**   *The following chromatographic experiments require the use of various organic solvents (alcohols, acetone, cyclohexane, toluene, and ethyl acetate), organic dyes, various phenolic compounds, and iodine. See HAZARD CATEGORIES 1, 3, 4 and 5, page 4.*

PROCEDURE

Separation of Dye Mixture by Column Chromatography

Prepare a small-scale chromatography column of the type illustrated in either Figure 7.1a or b. The column illustrated in Figure 7.1a is prepared by stoppering one end of a 20-cm length of glass tubing whose inner diameter is about 10 mm with a rubber stopper through which a medicine dropper has been inserted. Place a length of flexible tubing over the tip of the medicine dropper and attach a screw clamp on the tubing to function as a stopcock. The column illustrated in Figure 7.1b is prepared from a commercially available chromatography column that is equipped with a removable plastic stopcock. Commercial columns come in various sizes with either fixed or removable glass or plastic stopcocks, the latter being preferred. This experiment works best with a column that is about 200 mm long, 13 mm in outer diameter, and about 10 mm in inner diameter. (Note 1) Support the column with at least one buret clamp, preferably two. Using a length of glass rod or tubing, tamp a plug of glass wool gently but firmly into the lower end of the column (not so firmly that flow is impeded). Cover the plug with a 1-cm layer of fine sand.

Prepare a slurry by mixing 6 g of chromatographic grade alumina (Note 2) with 30 mL of 90% isopropyl alcohol in a small beaker. Fill the column about half full with 90% isopropyl alcohol, place a 250-mL Erlenmeyer flask under the column and open the stopcock (or pinch clamp). In the next step most students find it helpful to use a small funnel, although some students find it easier to work without one. Working slowly and carefully to avoid overflow from the top of the column, pour the slurry into the column using a micro-spatula to constantly mix the alumina with the isopropyl alcohol and scrape the slurry into the column while the excess solvent is escaping through the stopcock. *Never allow the column to be drawn dry of solvent before, during, or after use.* Add the slurry slowly so that the alumina settles through the solvent to form a nice, uniform column of packing. You may add additional isopropyl alcohol (from the Erlenmeyer flask) to the alumina in the beaker if necessary. Keep adding the slurry until the height of the alumina column above the sand is 6 cm. (Note 3) The flow through the column should slow quite a bit as the height of the column increases, but it should not stop altogether. Rinse the particles of alumina from the walls of the column with a little 90% isopropyl alcohol. Cover the top of the packing with a 0.5-cm layer of fine sand added through the solvent. (Note 4) Allow the column to drain until the last of the isopropyl alcohol just reaches the top of the upper sand layer; then close the stopcock.

To a 10-mL beaker add 0.5 mL of 90% isopropyl alcohol and 3–4 drops of a dye mixture prepared by dissolving 20 mg each of Rhodamine B and methylene blue

Figure 7.1   Simple column chromatography apparatus.

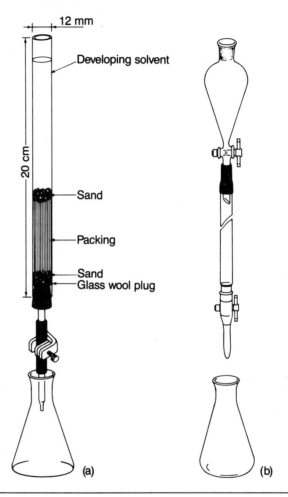

in 20 mL of 95% ethyl alcohol. Pick up this mixture in a long-stemmed Pasteur pipet, insert the pipet carefully into the column with the tip of the pipet just above the sand, and add the dye solution evenly to the top of the column. Open the stopcock and allow the dye to be drawn into the absorbent; then immediately fill the column with the developing solvent prepared by mixing 9 volumes of isopropyl alcohol with 1 volume of water. The column will develop slowly and steadily if you continue to replenish the solvent to maintain a sufficient head of liquid in the column; however, it is more convenient to fit the top of the column with a small separatory funnel as illustrated in Figure 7.1b. (Note 4) This provides an ample reservoir of solvent such that you can leave the column unattended for short periods of time. Continue to elute the dye mixture until each dye has been separated and received as a colored solution in the Erlenmeyer flask. If your instructor so directs, you may collect the two dyes in separate 50-mL flasks. When the second dye has made its exit from the column, close the stopcock and place a small cork in the top of the column. In this way your column can be kept for further use. Empty the flask(s) containing the eluted dye solutions into the receptacle provided for this purpose. Dye stains on the glassware can be removed with 95% ethanol.

PART II Paper Chromatography

A mixture to be resolved by paper chromatography is placed as a small spot on one end of a strip or sheet of paper and solvent is allowed to move by capillary action through the spot and up the paper. In this case, the water adsorbed on the cellulose of the paper is the stationary phase. The ratio of the distance traveled by a compound to that traveled by the solvent is called the *Rf* value.

$$Rf = \frac{\text{distance the compound traveled}}{\text{distance the solvent traveled}}$$

The *Rf* value of a compound is a characteristic of the compound and the solvent used and serves to identify each component of the mixture.

The following experiment illustrates how paper chromatography can be used to separate colored mixtures. The separation and identification of colorless compounds by paper chromatography is illustrated in procedure H, Experiment 20.

PROCEDURE   **The Separation of Food Colors by Paper Chromatography**

If your instructor does not provide micropipettes, prepare several from open-ended melting point tubes. Holding the melting point tube in both hands (from above), heat the tube in a small flame while rotating the tube in the fingers. When the center is soft, draw out the tube about 2–3 inches such that the drawn section is about 0.5 mm in diameter. Make two micropipettes by scratching and breaking the hair-like capillary about 0.5 to 0.75 in away from the thicker portion. The bore of the micropipette should be small enough that when it is touched to the surface of a solution or a liquid, a 1-cm column of liquid will be retained without a tendency for droplets to form.

Your instructor will provide you with some type of container or jar to use as a developing chamber. Any clear-glass, wide-mouthed bottle or jar that is about 15 cm deep will suffice. (Note 5) If the mouth of the jar is at least 70 mm in inner diameter, you may cut a 20 × 20 cm sheet of Whatman chromatography paper into three equal pieces (6.66 × 20 cm), which may be hung by a wire in the jar and used with up to four samples. Jars with smaller necks may also be used by rolling the paper into a partial tube. If jars are not available, use a tall beaker. Usually, if the beaker is 16 cm in height, its width will be such that the paper need not be rolled but may be hung from a wire, as for the wide-mouthed jar. Both techniques are described.

Figure 7.2 Spotting for paper or thin layer chromatography. The rule permits holding the paper or film securely. The metric scale on the rule allows you to space your spots uniformly.

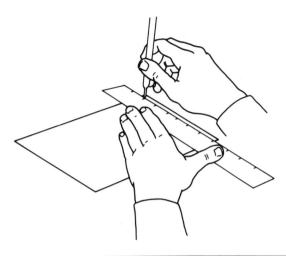

When the diameter of the jar is smaller than 65 mm, proceed as follows: Cut a rectangular piece of Whatman chromatography paper measuring 10 × 16 cm (see report form, p. 105). Draw a light *pencil* line about 1.5 cm from, and parallel to, the edge of the shorter side. Make four pencil marks on the bottom line at intervals of 2.5 cm and identify each mark as *R* (red), *B* (blue), *Y* (yellow), and *G* (green). Using separate micropipettes for each sample, make, in the above order, small (1.5-mm) spots of food colors on each mark (Fig. 7.2). Allow the spots to dry. Roll the paper sufficiently to pass through the neck of the developing jar. With the lower edge just touching the bottom of the jar, mark the level of the top of the jar's rim on the paper. Trim the paper to fit flush with the top and replace the paper in the jar. Fasten the paper to the rim with transparent tape. The two vertical edges of the paper must be parallel. The paper must not touch the walls of the jar. (See Fig. 7.3.) Dilute 10 mL of isopropyl alcohol with 5 mL of water in a small beaker and slowly transfer this solution to the bottom of the jar by pouring it through a long-stemmed funnel. Avoid splashing, and keep the level of the liquid below the spots on the paper. Cover the mouth of the jar tightly with aluminum foil and allow the chromatogram to develop for about 30–45 minutes. Remove the paper carefully but quickly from the jar and mark the upper edge of the solvent front before it evaporates. Attach the dry developed chromatogram to your report form.

If the mouth of the jar (or beaker) is at least 70 mm in diameter, you may proceed as follows: Cut a rectangular piece of Whatman chromatography paper whose length is 20 cm and whose width is 5–10 mm less that the diameter of the mouth of the developing jar. Insert the paper in the jar and mark the level of the top of the jar's

Figure 7.3   Paper chromatography assembly. (a) The filter paper is secured in the jar with tape. Lower edge of paper rests on bottom of jar. (b) Filter paper fitted with wire hanger.

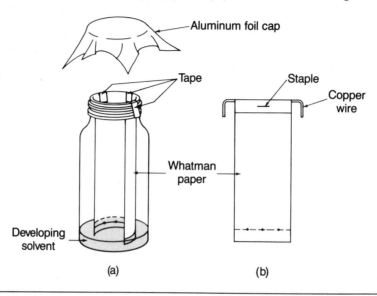

rim on the paper. Fold the paper at this mark over a length of stout copper wire bent as shown in Figure 7.3. Staple or tape the paper over the wire, trimming off excess paper. Draw a light *pencil* line about 1.5 cm from, and parallel to, the bottom of the paper. Make four pencil marks on the bottom line at intervals of 1.2 cm and identify each mark as *R* (*red*), *B* (blue), *Y* (yellow), and *G* (green). Using separate micro-pipettes for each sample, make, in the above order, small (1.5-mm) spots of food colors on each mark (Fig. 7.2). Allow the spots to dry. Dilute 10 mL of isopropyl alcohol with 5 mL of water in a small beaker, pour this solution into the jar, and swirl the solution around. Carefully lower the paper into the very center of the jar, making sure that the spots remain above the liquid level. Holding the paper and copper wire centered, cover the mouth of the jar tightly with aluminum foil and allow the chromatogram to develop for about 30–45 minutes. Remove the paper carefully but quickly from the jar and mark the upper edge of the solvent front before it evaporates. Attach the dry, developed chromatogram to your report form.

PART III   Thin Layer Chromatography

Thin layer chromatography, frequently referred to as TLC, is a technique that makes possible a rapid chromatographic separation on thin layers of adsorbent. In practice, a film of approximately 0.3 mm thickness of alumina, silica gel, or other substrate containing a small amount of binding material (usually calcium sulfate) is applied as a slurry to one side of a clean glass plate or plastic film. The slurry then is dried either at room temperature or by heating. Samples of mixtures to be separated are applied as small spots near the bottom of the plate or film, and the chromatogram is developed by following a procedure similar to that followed in paper chromatography. The plate or film is placed in an upright position in a chamber containing a small volume of developing solvent. As the solvent ascends the thin layer by capillary action, components of the mixture become separated and appear as distinct colored spots if the mixture is one of colored substances. If the mixture is composed of colorless substances, individual components may be rendered visible either by the use of ultra-violet light or by chemical means. Unlike paper, the thin mineral layer on the surface of the plate or film can be treated with a variety of reagents, including those of a corrosive nature. By marking the solvent front and measuring the distance traveled by each component, *Rf* values for individual compounds may be determined as in paper chromatography. In addition to being rapid, the method is very sensitive. Coated glass chromatoplates and coated plastic film for chromatography are available commercially.

PROCEDURE

The experiments in this manual were developed using Eastman Kodak TLC coated film (#13179 or #13181 [with fluorescent indicator]). There are several other good brands of TLC coated film available, and each of these has its own characteristics, its own advantages and disadvantages. If you use a different brand of TLC film, your observations may differ somewhat from those described here.

Eastman Kodak TLC coated film is provided as 10×10 cm sheets, which may be cut (with a paper cutter or, carefully, with heavy duty scissors) into 2.5×10 cm or 5×10 cm strips, which we will call a TLC plate. Unless these plates are carefully stored away from moisture, they may require activation (drying) in the oven at 105° for one hour for Part B.[1] Before using a plate, carefully draw a *very light pencil* line parallel to a short side of the plate about 1–1.5 cm from that side. Try not to scratch the coating. The surface coating of Eastman Kodak TLC coated film is more sturdy than most, but it is still fragile, particularly when damp with solvent. Up to three sample spots may be applied to a 2.5×10 cm plate; up to six sample spots may be applied to a 5×10 cm plate.

If TLC coated film is not available, you may make coated plates by the following procedure.

A. Coating Microscope Slides for TLC by Spreading the Adsorbent

Place fifteen to twenty-four 75×25 mm microscope slides of uniform thickness in two or three rows, side by side, on a smooth flat surface. (Note 2) Fasten the slides with plastic electrician's tape, allowing approximately one-eighth inch of tape to lap over the long edge of each outside row (see Fig. 7.4). In a 50-mL beaker prepare a slurry by mixing 5 g of silica gel[2] and 13 mL of water to a smooth creamy consistency

---

[1] Instructor's Guide.

[2] Ibid.

Figure 7.4   Preparing a number of microscope slides for TLC by spreading adsorbent in a thin layer with a glass rod.

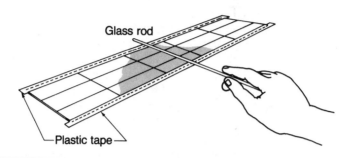

and transfer it to the slides. Spread the slurry over the slides in a smooth uniform layer by drawing a glass rod back and forth along the tape "rails." Remove the tape, place coated plates on clean pieces of paper toweling, and dry in an oven at 105° or over a hot plate. Store dried plates in a desiccator.

B.   The Separation of Dye Mixtures by TLC

Depending on the sample materials provided to you, you may carry out 1 or 2 below on a 2.5 × 10 cm or 5 × 10 cm plate, depending on the total number of spots to be applied.

1. In a 400-mL beaker place about 7–8 mL of a developing solvent mixture prepared from 5 parts by volume of 1-butanol, 5 parts of ethyl acetate, 5 parts of absolute ethanol, and 2 parts of water (for example, 5 mL each of the three organic solvents and 2 mL of distilled water should make enough solvent for two beakers). Volumes may be measured with a graduate cylinder, but try to be as accurate as possible because variations in the amount of water will affect the *Rf* values. Cover the top of the beaker tightly with a piece of aluminum foil to allow a solvent atmosphere to form while you are preparing your TLC plate. Using a micropipette, apply two small spots (1.0-mm diameter) of a previously prepared dye mixture (Note 6) on the sampling line on a TLC plate. Allow both spots to dry; then respot one of them. Allow the spot to dry again, remove the aluminum foil, and place the film carefully into the beaker, making certain that the sampling line is above the level of the liquid. Recover the beaker with the aluminum foil and observe the action on the plate. When the solvent has ascended to within about 1.5 cm of the top of the plate (approximately 30–45 minutes), mark the upper limit of the solvent front with a pencil, remove the plate, and allow it to dry. You should see three to four spots, depending on the dye mixture provided: red and blue spots close together, a pale orange spot at a higher *Rf* value, and a somewhat harder to see yellow spot near the sampling line. The dye solutions are deliberately made very dilute to give the best resolution with a minimum of

---

Figure 7.5   Calculation of *Rf* values.

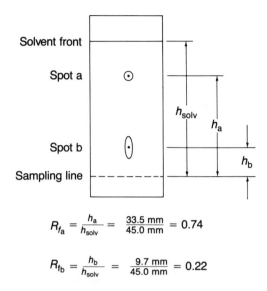

$$R_{f_a} = \frac{h_a}{h_{solv}} = \frac{33.5 \text{ mm}}{45.0 \text{ mm}} = 0.74$$

$$R_{f_b} = \frac{h_b}{h_{solv}} = \frac{9.7 \text{ mm}}{45.0 \text{ mm}} = 0.22$$

---

tailing. Measure the distance from the midpoint of each colored spot to the sampling line, and calculate *Rf* values for each dye used in the mixture (Fig. 7.5). Record these values on your report form and compare them with the average values given.

2. This procedure is exactly the same as that described in 1 above; however, the dye mixture will be taken from a Pilot Razor Point pen or a Flair Ultra Fine pen (preferably a red pen, but other colors may be used also). Make two spots with a given color and brand of pen, one about 0.5 mm and the other about 1 mm in diameter. *Practice first!* Some pens give up their ink more readily than others of the same type. After development with the solvent mixture and procedure described above, you should see three or four spots (not necessarily all well resolved) for either pen type. One of the spots from the Flair red pen is best seen with Eastman Kodak TLC coated film #13181 and the aid of an ultra-violet light designed for this purpose. **Caution!** *Do not look directly into any ultra-violet light source. It can damage your eyes.* This spot should fluoresce strongly under the ultra-violet lamp. If you have analyzed both a Flair and a Pilot red pen, record the differences and similarities in the dyes used in these pens. If you have time, you can examine other pens using the same technique.

C.   The Separation of Colorless Compounds by TLC

In a 400-mL beaker place about 6 mL of a developing solvent mixture prepared from 3 parts by volume of toluene, 1 part of ether, 1 part of 90% ethanol, and 1 part of acetone (for example, 1 mL each of the last three solvents and 3 mL of toluene).

Volumes may be measured with a graduate cylinder but try to be as accurate as possible. Cover the top of the beaker tightly with a piece of aluminum foil to allow a solvent atmosphere to develop while you are preparing your TLC plate. Using a micropipette, apply four small spots (1.0 mm diameter) of separate samples of phenol, catechol, resorcinol, and pyrogallol on a 5 × 10 cm TLC plate. Make, as a fifth spot, a composite of all four compounds. Prepare the samples by dissolving 30 mg of each phenolic compound in 1.0 mL of ethanol. Allow the spots to dry, then remove the aluminum foil and place the film carefully into the beaker, making certain that the sampling line is above the level of the liquid. Recover the beaker with the aluminum foil and observe the action on the plate. When the solvent has ascended to within 1.0–1.5 cm of the top of the plate (approximately 30 minutes), mark the upper limit of the solvent front with a pencil, remove the plate, and allow it to dry. Transfer the dry plate to a second clean, dry 400-mL beaker and add a small crystal (pinhead size) of iodine. Cover the beaker with aluminum foil and warm the bottom on an electric hot plate or over a low flame for a few seconds. Iodine vapor absorbed by the phenolic compounds will reveal the location of each. Carefully outline lightly with pencil each darkened oval area, as the color may fade completely on standing. Measure the distance from the midpoint of each spot on the developed chromatogram to the sampling line. Determine the *Rf* values (Fig. 7.5). Record these on your report form and compare them with the average values given.

PART IV   Gas Chromatography

Gas chromatography or, more specifically, gas-liquid (phase) chromatography (GLC or GLPC), is a technique employed for the separation and analysis of a mixture of gases, liquids, and even solids if the latter can be vaporized and have an appreciable vapor pressure at temperatures above 100°. To analyze a mixture by GLC, a small volume of the mixture is injected into a heated, packed column and the vaporized components are carried through the column by an inert *carrier* gas such as helium or nitrogen.

The separation of a mixture by GLC is based on the same principles as are separations by the chromatographic techniques described in Parts I, II, and III of this experiment, in that the components of the gaseous mixture being carried through the column also become partitioned between a stationary phase and a mobile phase. However, the packed column used in a gas chromatograph, unlike that used in gravity flow liquid-liquid chromatography, need be neither straight nor vertical. The column employed in GLC is often made of a coiled metal tubing of small diameter and is packed with a finely divided, inert, solid material, on the surface of which is adsorbed a liquid that is stable and nonvolatile. The former is called the support and the latter the stationary phase. The carrier gas acts as the mobile phase and transports the sample components from the injection port to a detector. The detector monitors the composition of the carrier gas stream as it leaves the column. The principal components of a gas chromatograph are shown schematically in Figure 7.6. A record of the detector response as a function of time is the chromatogram. The chromatogram consists of a base line corresponding to emergence of pure carrier gas and

Figure 7.6   Schematic diagram of a gas chromatograph.

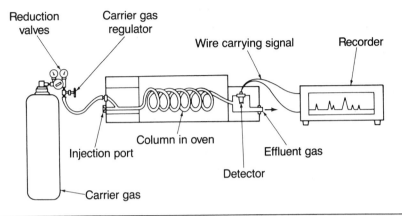

of peaks corresponding to emergence of carrier gas plus sample components. The detector produces a response that is proportional to the concentration of substance in the effluent from the column. A peak area, therefore, is proportional to the amount of a given constituent in the mixture. However, the detector's response may not be exactly the same for two equal concentrations of two different substances. Therefore, peak areas are not necessarily the same for equal amounts of different substances but rather are functions of the individual substances. The composition of a mixture, however, can be calculated from measurements of peak-area ratios. Peak areas may be obtained by use of a mechanical or electronic integrator, by height and width measurements, and even by the simple expedient of cutting out the area and weighing the chart paper on a sensitive balance.

In this experiment we will analyze a mixture of 2-propanol (bp 82.3°) and cyclohexane (bp 81°). As you may see from their boiling points, it would be difficult to separate a mixture of these two compounds by fractional distillation. However, they are quite easily separated and their percentage composition in a mixture can be determined using GLC.

Injection Technique

Good results in any gas chromatography experiment are dependent to a great degree upon the injection technique of the experimenter. You must exercise care to reproduce both the technique and the sample size as accurately as possible. The syringe you use should be rinsed several times when a new sample is to be analyzed. To rinse the syringe, fill it with sample, expel the contents, and wipe the needle dry with cleansing tissue. To analyze a sample, fill the syringe with a volume in excess of that required, return the plunger to the *exact* volume needed, then retract the plunger to draw in air and wipe the needle dry. (Note 4) To inject the sample, center the needle on the septum of the injection block, guiding it between the thumb and forefinger to prevent the possibility of bending the needle when it is inserted through the septum. **Caution!** *The block is hot!* Insert the needle rapidly, inject the sample, and immediately withdraw the needle from the injection port.

PROCEDURE

Although no general operating procedure can be given for every type of gas chromatographic separation, the procedure that follows is relatively basic and can be applied to any GLC instrument if modified to that instrument's operating instructions. The instruments used in our student laboratories are student instruments manufactured by Gow-Mac (Series 150) and Carle Instruments. However, there are many other fine brands of GLC instruments, any of which can be used in this experiment. We used both the DC-200 on Chromosorb P column that comes with the Gow-Mac instrument or a 6-foot long, 1/8-in diameter column packed with 8% SF-96 on Chromosorb WHP (80/150 mesh) in a Carle instrument. Other medium-polarity, silicone oil stationary phases should work equally well. (Note 7)

Obtain from your instructor 2-mL samples of 2-propanol (isopropyl alcohol), a standard mixture of cyclohexane and 2-propanol, and an unknown mixture of these two compounds. Adjust the instrument setting to obtain the flow rate that your instructor informs you is the best for your instrument and column. Balance the recorder. Load approximately 0.5 μl of 2-propanol and 5.0 μl of air into the syringe and inject the sample into the column. Mark the chart paper at this point. When the 2-propanol is completely eluted, that is, when the peak has reached a maximum and returned to the base line, inject a 0.5-μl sample of the standard mixture. It may be necessary here to adjust the recorder attenuation to keep the pen on scale. After you have eluted both components of the known mixture, inject a 0.5-μl sample of your unknown. After your unknown sample has been eluted, tear off your portion of chart paper and write your name on it. Measure the distance (or note the time) between the air peak and the 2-propanol peak on the 2-propanol chromatogram. This distance in time units is called the **retention time** (t_p) for 2-propanol and should be a constant for a given column operated under exactly the same conditions. Measure the retention times of the peaks for the standard sample and also the unknown mixture. With this information, identify the peaks that are those of the cyclohexane–2-propanol standard sample (C_s and P_s), as well as those of the cyclohexane–2-propanol unknown mixture (C_u and P_u).

Compute the area (A) under each peak in the chromatogram by carefully measuring the peak height and its width at the peak's half-height and multiplying these two values (see Fig. 7.7).

Assume that the area under the peak is proportional to the weight (wt) of substance in the sample. Then $A_P = K_P Wt_P$ and $A_C = K_C Wt_C$.

K_P and K_C are constants; however, $K_P \neq K_C$ because the detector response will probably not be the same for equal concentrations of 2-propanol and cyclohexane.

$$Wt\%C = \frac{Wt_C \, 100}{Wt_C + Wt_P} = \frac{100(A_C/K_C)}{A_C/K_C + A_P/K_P} = \frac{100(A_C/A_P)(K_P/K_C)}{1 + (A_C/A_P)(K_P/K_C)}$$

Figure 7.7   Measurement of the peak area.

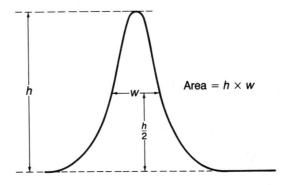

Area = $h \times w$

We need only to know a value for the K_P/K_C ratio, not the value of either K_P or K_C. This constant ratio (K_r) can be determined from the chromatogram of the standard sample using the formula

$$\frac{K_P}{K_C} = \frac{\text{Wt\%C}_s}{\text{Wt\%P}_s(A_C/A_P)_s} = K_r$$

The weight percentage of cyclohexane in the unknown mixture can then be determined from

$$\text{Wt\%C}_u = \frac{100\left(\dfrac{A_C}{A_P}\right)_u K_r}{1 + \left(\dfrac{A_C}{A_P}\right)_u K_r} = \frac{100K_r}{\left(\dfrac{A_P}{A_C}\right)_u + K_r}$$

and

$$\text{Wt\%P}_u = 100 - \text{Wt\%C}_u$$

Note 1   The Kontes K420530, Size 0212, column has the dimensions 13 mm o.d. and 10.5 mm i.d. × 200 mm long and has a removable plastic stopcock. The Corning 2145 column is similar but longer and has a fixed stopcock. If a longer column is used, add the sample mixture by using a Pasteur pipet and a length of small bore tubing so that sample does not splash on the inner walls of the column. A 25-mL buret with a removable plastic stopcock may be used in place of a regular chromatography column, again using a length of glass tubing to direct the placement of the sample.

Note 2   There is a variety of grades of alumina sold for use in chromatography. We used Aldrich Chemical Company activated, neutral Brockmann I alumina (catalog number 19,977-4).

Note 3   Generally, the height of the alumina column should be about ten times its diameter. A shorter column is used here to eliminate the necessity of applying either suction to the outlet or pressure to the inlet of the column to get a reasonable flow rate, as is required with longer columns.

Note 4   A small circle of filter paper (slightly smaller than the inner diameter of the column) is sometimes added before the layer of sand.

Note 5   A *wide-mouthed* (not the regular size) 1-quart mason jar of the type used in home canning works very well. The mouth on these jars is about 70–75 mm in inner diameter, and the jar is about 165 cm deep. Fisher Scientific Company sells 32-oz, screw-cap bottles with 89-mm necks (catalog numbers, 03-320-3M and 03-320-7F).

Note 6   The dye mixture is prepared by weighing 40 mg of **Rhodamine B,** 50 mg of **Bromphenol Blue,** 50 mg of **Orange II (Acid Orange 7),** and 60 mg of **Tartrazine (Acid Yellow 23)** into a 50-ml Erlenmeyer flask and adding 20 mL of 50% aqueous ethanol. If desired, 40 mg of fluorescein may also be added; however, the separation will be somewhat less complete. These concentrations give spots that are fairly easy to see, but, if desired for simplicity, 40 mg of each dye may be used at the expense of slightly weaker spots for the yellow and orange dyes. Make up fresh dye mixture in quantities no larger than needed, as some of the dyes fade on standing in aqueous ethanol. See the Instructor's Guide for a discussion of such issues as the effects of pH, variation in the composition of the developing solvent mixture, and type of TLC plate.

Note 7   See the Instructor's Guide for a discussion of instrument settings, alternative combinations of liquids, and alternative column packings. ***Caution!*** *Safety considerations require that samples injected into the hot inlet system of a gas chromatograph be free of thermally unstable substances.* ***Read Note 1, p. 57.***

PART I   Column Chromatography

Describe briefly your observations during this part of the experiment.

Name _____ Section _____ Date _____

PART II Paper Chromatography

The Separation of Food Colors

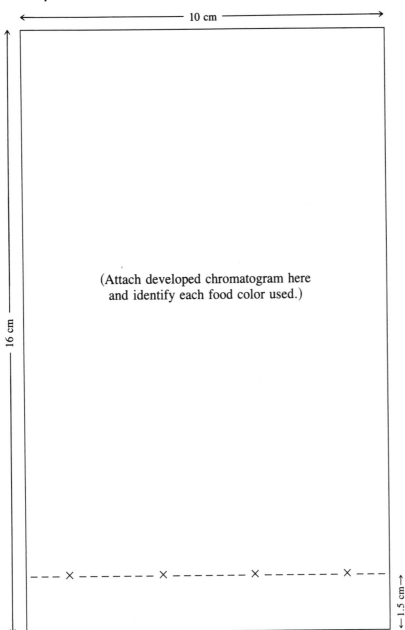

(Attach developed chromatogram here
and identify each food color used.)

PART III   Thin Layer Chromatography

A.   The Separation of Dye Mixtures

Dye	R_f values (average)	R_f values (experimental)
Tartrazine	0.21	
Rhodamine B	0.52	
Bromphenol Blue	0.59	
Orange II	0.71	

If you used coated microscope slides for TLC dye separation, indicate the location of each dye in your mixture by sketching the developed chromatogram in the outline at the right. If you used coated plastic film, staple your developed chromatogram in the space below.

Name _____ Section _____ Date _____

PART III   Thin Layer Chromatography

B. The Separation of Dye Mixtures from Pens

Pen:		Pen:	
Color	*Rf*	Color	*Rf*

If you used coated microscope slides for TLC dye separation, indicate the location of each dye in your mixture by sketching the developed chromatogram in the outline at the right. If you used coated plastic film, staple your developed chromatogram in the space below.

Name   Section   Date

PART III   Thin Layer Chromatography

C.   The Separation of Colorless Compounds

Pure Compounds	*Rf* values (average)	*Rf* values (experimental)
Phenol	0.85	
Catechol	0.75	
Resorcinol	0.72	
Pyrogallol	0.63	

Mixture	*Rf* values (average	*Rf* values (experimental)
Phenol	0.85	
Catechol	0.72	
Resorcinol	0.77	
Pyrogallol	0.53	

If you used coated microscope slides for TLC separation of phenolic compounds, indicate the location of each compound by sketching the developed chromatogram in the outlined space. If you used coated plastic film, staple your developed chromatogram in the space below.

Name   Section   Date

PART IV   GLC Chromatography

The Analysis of an Unknown Mixture of Cyclohexane and 2-Propanol

The chromatographic record of your analysis will be a necessary part of the report for Part IV of the chromatography experiments. Include it along with the following information:

Sample Number _____

$Wt\%C_s$* _____ ; $Wt\%P_s$* _____

t_C _____ ; t_P _____

A_{Cs} _____ ; A_{Ps} _____

A_{Cu} _____ ; A_{Pu} _____

K_r = _____

$Wt\%C_u$ _____ ; $Wt\%P_u$ _____

_____
*Percentages to be furnished by the instructor.

Reactions of Hydrocarbons

In this experiment you will carry out some simple chemical tests on the four families of hydrocarbons: the alkanes, the alkenes, the alkynes, and the arenes (aromatic hydrocarbons). There are wide variations in the reactivities of the various classes of hydrocarbons toward the reagents used in this experiment; therefore, do not expect that each hydrocarbon studied will react with all of the reagents.

PART I   Aliphatic Hydrocarbons

The aliphatic hydrocarbons are divided into three classes: (1) the **alkanes,** (2) the **alkenes,** and (3) the **alkynes.** The alkanes have only single bonds in the molecule and are said to be _saturated_ hydrocarbons. They are relatively inert. The alkenes have at least one double bond in the molecule, and the alkynes have at least one triple bond. Both the alkenes and the alkynes are said to be _unsaturated_ hydrocarbons, and both are relatively reactive.

> **Precautions.**   _This experiment requires the use of volatile hydrocarbons, bromine, sulfuric acid, nitric acid, sodium metal, sodium hydroxide, and ammonium hydroxide. See HAZARD CATEGORIES 1, 2, and 3, page 4._

PROCEDURE   In this and other experiments in this text, when dropwise addition of reagents is specified, use a Pasteur pipet (rather than a conventional medicine dropper) unless your instructor or this manual directs you to do otherwise. The droplets from a Pasteur pipet are generally smaller than those from a typical medicine dropper and permit slower, more controlled addition. You may wish. to calibrate the pipets available to you by counting drops into a 10-ml graduate cylinder.

   After each test is completed, dispose of the reactants in a safe manner. Pour mixtures containing sulfuric acid _cautiously_ into a large volume of water while stirring; then the diluted acid may be discarded down the sink drain. Water-soluble mixtures (aqueous and ethanolic solutions) may be diluted with much water and discarded down the sink drain. Mixtures containing methylene chloride or toluene should be disposed of as your instructor directs.

A. Alkanes

Obtain 5–6 mL of an alkane or alkane mixture from the bottle labeled "Alkane" on the side shelf. This bottle will contain a moderately pure sample of an alkane or mixture of alkanes commonly used as a solvent in the organic chemistry laboratory. Typically, for this experiment, it will be a hexane, heptane, or a mixture of these, boiling in the range of about 69–98°. (Note 1) Carry out the following reactions in *dry* test tubes.

(a) Bromine in Methylene Chloride[1]

Alkanes react slowly or not at all with bromine *in the dark* but in the presence of light may react fairly rapidly according to the equation

$$R\text{---}H + Br_2 \longrightarrow R\text{---}Br + HBr$$

To each of two 2-mL portions of Alkane in separate test tubes add 0.5 mL (about 10 drops) of bromine in methylene chloride solution. Shake the tubes well, place one tube in the desk in the dark and expose the other to a bright fluorescent lamp or incandescent bulb, or better still, to direct sunlight for a period of about 15 minutes. Compare the color of the two tubes. Blow your breath *gently* across the mouth of each tube and observe the result. If hydrogen bromide is present, it will combine with the moisture of the breath to form a faint cloud.

(b) Aqueous Potassium Permanganate (Baeyer's Test)

Under normal conditions the alkanes are stable (that is, unreactive) toward chemical oxidizing reagents.

Caution! *Potassium permanganate is a strong oxidant and will stain the skin. Wash off any spills immediately with large amounts of water.*

To each of two test tubes add 2 mL of 95% ethanol. To the first add 2 drops of Alkane. The second test tube will serve as a blank. To each test tube add 10 drops of 0.5% aqueous potassium permanganate solution. Shake the test tubes for several minutes to allow time for any reaction to take place and carefully observe both for any color change or appearance of a precipitate. If a color change occurs, add a second 10 drops of 0.5% aqueous potassium permanganate solution to *both* test

---

[1]**Note to Instructor.** Methylene chloride is not stable to bromine, but will serve as a reagent in the unsaturation test if prepared and stored in a brown glass-stoppered bottle. It is best prepared in small volumes shortly before needed and disposed of after a week or two.

tubes, and continue to shake to mix the reactants. A positive test is the disappearance of the permanganate color or the appearance of a brown solution or precipitate. Record your observations on the report form.

(c) Sulfuric Acid

Alkanes do not react with sulfuric acid under normal conditions.

To 2 mL of concentrated sulfuric acid add 10 drops of Alkane and shake vigorously. **Caution!** *Do not use your thumb as a stopper.* Note if heat is evolved or if the alkane appears to dissolve.

B. Alkenes

Obtain 5–6 mL of cyclohexene from the side shelf and carry out the following reactions in dry test tubes.

Cyclohexene is a cyclic alkene having the structure [ ]. The reactions of cyclohexene are the same as those of the acyclic or "straight" chain alkenes. It is used in this experiment because it is readily available in pure form and because it has a fairly high boiling point and will not evaporate during your tests.

(a) Bromine in Methylene Chloride

Bromine *adds* rapidly to unsaturated hydrocarbons to give organic halogen compounds.

$$R\!-\!CH\!=\!CH\!-\!R + Br_2 \longrightarrow R\!-\!\overset{\displaystyle H}{\underset{\displaystyle Br}{C}}\!-\!\overset{\displaystyle H}{\underset{\displaystyle Br}{C}}\!-\!R$$

Inasmuch as halogen derivatives of hydrocarbons are colorless liquids or solids, decolorization of a solution of bromine in methylene chloride may be used as a *test for unsaturation*, provided that no other functional group is present that reacts with bromine.

To 1 mL of cyclohexene add 5 drops of 2% bromine in methylene chloride solution, shake the tube, and observe the result. Keep adding 5-drop increments of bromine in methylene chloride solution until a total of 50 drops has been added. Does the orange color of the bromine in methylene chloride solution persist, or does your sample continue to decolorize the reagent? Again, blow your breath gently across the mouth of the test tube and observe.

(b) Aqueous Potassium Permanganate (Baeyer's Test)

In the absence of other easily oxidized groups, alkenes react with neutral or alkaline potassium permanganate solution to form glycols according to the reaction

$$3 \text{ R—CH}=\text{CH—R} + 2 \text{ KMnO}_4 + 4 \text{ H}_2\text{O} \longrightarrow$$
(purple)

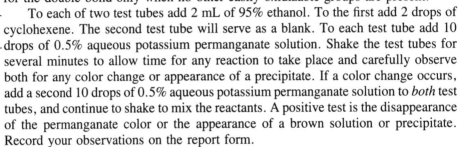

$$3 \text{ R—}\underset{\underset{\text{H}}{|}}{\overset{\overset{\text{OH}}{|}}{\text{C}}}\text{—}\underset{\underset{\text{H}}{|}}{\overset{\overset{\text{OH}}{|}}{\text{C}}}\text{—R} + 2 \text{ MnO}_2 + 2 \text{ KOH}$$

(a glycol)   (brown)

This reaction is the basis for the Baeyer's test for a double bond. Evidence of the reaction with permanganate is the change in color from the purple of the aqueous permanganate solution to the brown of the manganese dioxide precipitate. The glycols produced by the reaction are colorless and water-soluble. This test is specific for the double bond only when no other easily oxidizable groups are present.

To each of two test tubes add 2 mL of 95% ethanol. To the first add 2 drops of cyclohexene. The second test tube will serve as a blank. To each test tube add 10 drops of 0.5% aqueous potassium permanganate solution. Shake the test tubes for several minutes to allow time for any reaction to take place and carefully observe both for any color change or appearance of a precipitate. If a color change occurs, add a second 10 drops of 0.5% aqueous potassium permanganate solution to *both* test tubes, and continue to shake to mix the reactants. A positive test is the disappearance of the permanganate color or the appearance of a brown solution or precipitate. Record your observations on the report form.

(c) Sulfuric Acid

Alkenes react with cold concentrated sulfuric acid to give alkyl hydrogen sulfates.

$$\text{R—CH}=\text{CH—R} + \text{HO—}\overset{\overset{\text{O}}{\uparrow}}{\underset{\downarrow}{\text{S}}}\text{—OH} \longrightarrow \text{R—}\underset{\underset{\text{H}}{|}}{\overset{\overset{\displaystyle\text{OH}}{\displaystyle|}}{\underset{\displaystyle|}{\text{O←S→O}}}}\text{—CH}_2\text{—R}$$

As the hydrogen sulfates are soluble in concentrated sulfuric acid, the observer sees only the alkene appearing to dissolve. Although the alkene appears to dissolve in the sulfuric acid, it should be noted that a reaction rather than true solution is involved.

To 2 mL of concentrated sulfuric acid add 1 mL of cyclohexene in small portions, shake after each addition, and observe the result. Report all your findings on the report sheet.

C. Alkynes

The number of commercially available alkynes, although increasing, is very limited, and some of these are rather expensive. In this experiment you will use 1-heptyne and 2-methyl-3-butyn-2-ol (a relatively inexpensive hydroxy alkyne). Use 1-heptyne prudently, as it is moderately expensive.

(a) Formation of Metallic Salts

1-Alkynes (terminal acetylenes) react readily with aqueous ammoniacal silver nitrate to give water-insoluble acetylides according to the following equation.

$$R\!-\!C\!\equiv\!C\!-\!H \;+\; Ag(NH_3)_2{}^+NO_3{}^- \;\longrightarrow$$
$$R\!-\!C\!\equiv\!C\!-\!Ag \;+\; NH_4{}^+NO_3{}^- \;+\; NH_3$$

The silver acetylide is a shock-sensitive explosive when dry; therefore, in this experiment you should decompose the salts by adding dilute nitric (or hydrochloric) acid at the end of the experiment.

Only alkynes that have at least one hydrogen on the triply bonded carbon atom show this reaction. Alkenes and alkanes do not react.

Step 1   In a test tube prepare an ammoniacal solution of silver nitrate by adding concentrated ammonium hydroxide *dropwise* to 5 mL of a 2% solution of silver nitrate. Add the ammonium hydroxide slowly until the brown precipitate of silver oxide that forms initially *just* dissolves. Pour half of the solution into a second test tube. Dilute the solution in one test tube with 2 mL of water and add 2–3 drops of 2-methyl-3-butyn-2-ol. Dilute the solution in the second test tube with an equal volume of 95% ethyl alcohol and add 2–3 drops of 1-heptyne. Observe and record the results in both test tubes. Collect the precipitates on the Hirsch funnel, being careful not to let the precipitates dry completely, for they are explosive when dry. Decompose the silver salts by treatment with dilute nitric acid making sure that all traces of the precipitates have dissolved before pouring the solutions down the drain.

Step 2   Place 5 mL of *anhydrous* toluene in a clean, dry test tube. Add to the toluene a sliver of clean sodium about 2–3 mm in diameter. Carefully observe the mixture for the evolution of tiny bubbles. If these form, the toluene is not anhydrous (or the test tube was not dry). When bubble formation has ceased, or nearly so, add 3–4 drops of 1-heptyne to the mixture and swirl gently to mix the liquids. Carefully observe the mixture for the evolution of tiny bubbles. Allow the tube to stand for 30 minutes (in the hood); then observe the surface of the sodium for evidence of salt formation. Destroy the unreacted sodium by adding 2–3 mL of 95% ethanol to the toluene solution. After all of the sodium has dissolved, pour the solution into the toluene waste receptacle. Write an equation for any observed reaction on your report sheet.

(b) Reaction with Bromine

To each of two test tubes add 2 mL of methylene chloride. To the first add 2 drops of 1-heptyne and to the second add 2 drops of 2-methyl-3-butyn-2-ol. Counting the drops, to each test tube add 2% bromine in methylene chloride dropwise, swirling to mix the reactants until the bromine color is no longer decolorized.

(c) Reaction with Potassium Permanganate

Although the reaction of alkynes with potassium pemanganate is more complex than that with alkenes, the observed results are the same. To each of three test tubes add 2 mL of 95% ethanol. To the first add 2 drops of 1-heptyne and to the second add 2 drops of 2-methyl-3-butyn-2-ol. The third test tube will serve as a blank. To each test tube add 10 drops of 0.5% aqueous potassium permanganate solution. Shake the test tubes for several minutes to allow time for any reaction to take place and carefully observe both for any color change or appearance of a precipitate. If a color change occurs, add a second 10 drops of 0.5% aqueous potassium permanganate solution to *each* test tube, and continue to shake to mix the reactants. A positive test is the disappearance of the permanganate color or the appearance of a brown solution or precipitate. Record your observations on the report form.

PART II   Aromatic Hydrocarbons

Aromatic hydrocarbons are those that contain one or more C_6 benzenoid structures.

The unsaturation of benzene is due to 6 *p* electrons (one from each carbon atom) in overlapping orbitals that form π bonds between carbon atoms in the ring. The π electrons are not fixed (localized) between any two carbon atoms, but rather are distributed so as to encompass all six carbon atoms within the ring. This overall distribution (delocalization) of electrons confers upon benzene and its derivatives a greater stability than if they were simply cyclic, conjugated trienes.

The reactions of the aromatic hydrocarbons, like those of the saturated hydrocarbons, are principally reactions of substitution.

The following experiments will serve to illustrate some of the properties and some of the more common substitution reactions of toluene—a methyl benzene.

PROCEDURE   A.   Solubility

⌐Test the solubility of toluene in water, ethanol, and hexane or petroleum ether using
└ 0.5-mL portions of toluene.

B.   Flammability

Ignite a few drops of toluene in a small evaporating dish **(HOOD!).** Repeat the
flammability test using alkane. Compare the luminosity of the flame and the amount
of soot formed in each case.

C.   Action of Sulfuric Acid (Sulfonation)

In a test tube add 1.0 mL of toluene to 8 mL of concentrated sulfuric acid and warm
in a water bath maintained at a temperature no higher than 50°. (**Caution!** *No
flames.*) Occasionally remove the tube from the bath, stopper tightly with a rubber
stopper, and shake (**Caution!** *Keep finger on stopper. Hold test tube away from
body*). Continue heating and shaking mixture until two layers no longer separate.
Cautiously pour one-half of the reaction mixture onto 10 g of ice. Transfer the
aqueous solution to a test tube and examine. Result? On your report form write the
equation for the reaction of toluene with concentrated sulfuric acid.

D.   Action of Nitric Acid (Nitration)

In a 50-mL Erlenmeyer flask prepare a nitrating mixture by cautiously adding 3 mL
of concentrated sulfuric acid to 2 mL of 8*M* nitric acid. Carefully add 0.5 mL of
toluene dropwise, shaking to mix. After several minutes pour the entire contents of
the flask onto 15–20 g of cracked ice. Note the heavy oil that separates. Write the
equation for this reaction.

E.   Action of Bromine (Bromination)

Caution! *One of the reaction products of this experiment is benzyl bromide—a
potent lachrymator. Therefore, carry out this procedure in a good fume hood and
provide a container in the hood as a receptacle for the disposal of reaction products.*

Place 1 mL of toluene into a 50-mL beaker and 1 mL of toluene in a test tube.
Introduce into each of the two samples 2–3 drops of bromine from the bromine buret
(HOOD!). To the test tube sample add a few iron filings or an iron tack. Place the
sample in a warm (65°) water bath for 30 minutes. Place the beaker sample directly
under a 100-watt incandescent lamp for 2–3 minutes. Blow your breath gently across
the mouth of the test tube and the beaker. Observe any color changes. In which
sample did a chemical change occur more rapidly? Write equations for all reactions
that take place.

F. Test for Unsaturation

To each of two test tubes add 2 mL of 95% ethanol. To the first add 2 drops of toluene. The second test tube will serve as a blank. To each test tube add 10 drops of 0.5% aqueous potassium permanganate solution. Shake the test tubes for several minutes to allow time for any reaction to take place, and carefully observe both for any color change or appearance of a precipitate. If a color change occurs, add a second 10 drops of 0.5% aqueous potassium permanganate solution to *both* test tubes, and continue to shake to mix the reactants. A positive test is the disappearance of the permanganate color or the appearance of a brown solution or precipitate. Record your observations on the report form.

Note 1 In previous editions of this manual we suggested the use of Skellysolve C (bp 88–98°) or of petroleum ether (alkanes of bp 35–36°). These are still suitable if available. However, current practice suggests that solvents of a narrow boiling point range are often preferred. Relatively inexpensive alkane solvents suitable for this experiment that meet the boiling point range criterion are hexane (bp 69°), hexanes (bp 68–72°), and heptane (bp 98°). The last of these is especially attractive for this experiment, based on cost and lesser volatility.

Name   *Section*   *Date*

PART I

Aliphatic Hydrocarbons

Reagents used	Alkane	Cyclohexene
Br_2 in CH_2Cl_2 (dark)		
Br_2 in CH_2Cl_2 (light)		
0.5% $KMnO_4$		
Conc. H_2SO_4		

Acetylene

Formation of Metallic Salts

(a) Write equations for the reactions of your alkynes with $AgNO_3$.

(b) What is the color of the precipitate in the above reaction? _____

(c) Write an equation for the reaction of 1-heptyne with sodium.

Reaction with Bromine

Write equations for the reactions of your alkynes with bromine.

Reaction with Potassium Permanganate

Was the color of permanganate discharged? Explain. _____

_____

PART II

Aromatic Hydrocarbons

Solubility of Toluene

Water	Ethanol	Hexane

Flammability of Toluene and Alkane

Describe the appearance of the flames. Toluene: _____

_____

Alkane: _____

Write an equation for the *complete* combustion of toluene.

Action of Concentrated Sulfuric Acid on Toluene

Write an equation for the preparation of toluenesulfonic acid.

Name   *Section*   *Date*

Action of Nitric Acid on Toluene

Write an equation for the nitration of toluene.

Action of Bromine on Toluene

Write an equation for the reaction of bromine with toluene in the presence of iron filings.

Write an equation for the reaction of bromine with toluene in the presence of bright light.

Action of Potassium Permanganate on Toluene

Is there any evidence of reaction?

1. What is the nature of the gas that evolves when an alkane reacts with bromine?

2. When potassium permanganate solution is decolorized by a sample of a saturated hydrocarbon what conclusions can you draw?

3. Which of the above reactions would you expect to take place with a sample of gasoline? With kerosene?

4. Why was it necessary to use anhydrous toluene in the reaction of 1-heptyne with sodium?

5. If you were to run the sodium test with 2-methyl-3-butyn-2-ol, would you expect more or less evolution of hydrogen gas? Write an equation to justify your answer.

6. A cylinder of compressed gas has no identifying marking. The gas is flammable and could possibly be either propane or acetylene. Devise a simple test that would identify the gas.

7. What physical property of aromatic hydrocarbons is greatly altered by the presence of a sulfonic acid group, $-SO_3H$, on the aromatic ring?

8. What common physical properties do nearly all nitro compounds share? What common chemical property?

9. In each of the Procedures C, D, and E of Part II more than one reaction product is possible. Write the structures and assign acceptable names to all possible products.

10. Draw the structures of T.N.T. and T.N.B. Which of these two high explosives is more easily made? Why?

11. What explanation would you offer if, in Procedure F of Part II, the color of permanganate had been discharged? If hot, alkaline $KMnO_4$ had been used on toluene, what would your product have been?

Steam Distillation:
The Isolation of (+)-Limonene
from Orange Peel

The distillation of a homogeneous binary mixture (Experiment 4-A) produces a vapor to which each component in the mixture contributes a part. The fraction contributed by each is equal to the mole fraction (N) of the component in the mixture multiplied by the value of the vapor pressure ($P°$) each component alone would exert at the boiling point of the mixture. This is a statement of Raoult's law, which in equation form appears in every beginning chemistry text as

$$P_{total} = P_A° N_A + P_B° N_B$$

The distillation of a heterogeneous binary mixture, on the other hand, produces a vapor mixture that is not dependent upon the mole fraction of each component present, but only upon the temperature. Thus, the total pressure at the boiling point of the heterogeneous mixture is simply the sum of the two separate vapor pressures, or

$$P_{total} = P_A° + P_B°$$

Thus, if a mixture is composed of water and a high-boiling, water-immiscible substance, the mixture will boil near but somewhat below 100°. Most of the distillate will be water, but a separable portion of it will be the immiscible component, usually in a high state of purity. The process just described, called **steam distillation,** is an excellent technique for the isolation of many water-insoluble compounds that are unstable at temperatures near their boiling points, or compounds difficult to isolate or purify in any other way.

Steam distillations may be carried out in one of two ways. In one method, called **direct steam distillation,** the compound to be separated and water are simply placed in a conventional distillation apparatus and the distillate is collected. The distillate, which appears as a heterogeneous mixture, may be either two liquid layers or a solid and water. A separation is easily effected by way of the separatory funnel or by a simple filtration. **Indirect steam distillations** are carried out by piping steam generated outside the distillation flask into a mixture of compound and water while the distillation flask itself is kept at a temperature near 100° in order to minimize a build-up of condensate. Apparatus for both methods are illustrated in Figure 9.1. In the present experiment you will isolate by way of direct steam distillation a sample of limonene from one of its many natural sources.

Figure 9.1 Assemblies for steam distillation. (a) Direct distillation. (b) Indirect steam distillation.

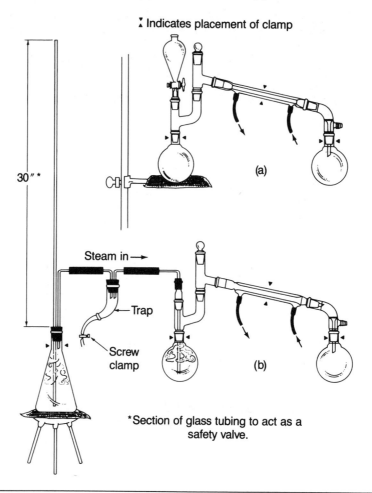

⟲ Indicates placement of clamp

30″ *

Steam in →

Trap

Screw clamp

(a)

(b)

*Section of glass tubing to act as a safety valve.

PROCEDURE Assemble an apparatus as illustrated in Figure 9.1a. Use a 500-mL flask as the distillation flask and a 250-mL flask as the receiver. Heat the distillation flask with either a burner and a ceramic fiber-centered wire gauze (as shown) or with a 500-mL heating mantle. (Note 1) If you use a heating mantle, support it in such a way that it can be removed quickly if necessary to control the distillation. An iron ring will provide satisfactory support for an ordinary glass fabric-covered mantle.

Next, with the help of a high-speed food blender, prepare a puree from the parings of four large navel oranges. Use sufficient water to produce approximately 500 mL of slurry. (Note 2) Place 300 mL of the orange peel slurry in the distillation flask and bring the mixture to a steady boil. Avoid violent boiling, for excessive frothing results and may carry small amounts of slurry directly into the condenser. Should this occur, the contents of the receiver must be returned to the distillation flask, the condenser flushed with fresh water, and the distillation resumed more judiciously. Periodically add 5–10 mL of water to the distillation flask from the separatory funnel to maintain a constant volume. Continue the distillation until about 125 mL of condensate has been collected. By this time you may see an oily layer on the surface of the collected distillate. If no oily layer appears, it may be necessary to collect a larger volume of condensate or to recharge the distillation flask with fresh slurry. When it is apparent that you have collected some limonene, stop the distillation and discard any water remaining in the separatory funnel. Transfer the distillate to the separatory funnel and separate the organic layer from the water. Store the limonene collected in a clean test tube over 10–12 granules of anhydrous calcium chloride. The yield of limonene will be 2–4 mL. After the limonene has dried, carry out the following tests. (Note 3)

1. Obtain an ir spectrum of your sample of limonene and compare it to that shown in Figure 9.2. Attach the ir spectrum to your report.

2. The structure of limonene shows the presence of two double bonds. Test for unsaturation by preparing a solution of 10 drops limonene in 1 mL of methylene chloride and carry out the bromine absorption test as described in Experiment 8, Part I-B.

3. Although your ir spectrum of limonene may appear to exactly duplicate that shown in Figure 9.2, you cannot be certain that your sample does not contain

---

Figure 9.2   Infrared spectrum of limonene (neat).

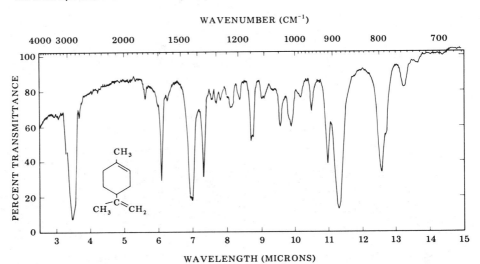

an impurity. To learn if you have a pure substance, analyze a sample of limonene using a gas chromatographic column. Follow the general procedure described in Experiment 7, Part IV. A column packed with Durapak Carbowax 400 and maintained at a temperature of 150° and a flow rate of 20 mL/minute will be satisfactory. Record the results of the GLC study on your report form. Also attach the chromatogram.

4. The structure of limonene further indicates that it is a chiral molecule because it possesses a center of asymmetry in a carbon atom bonded to four dissimilar groups. Inasmuch as limonene has its origin in a natural source, we could reasonably expect it to be optically active, one of the compounds in the current chiral pool. One-half of the class should make a solution of 0.75 g of limonene in 25 mL of absolute ethanol and the other half of the class should make a solution of 1.5 g of limonene in 25 mL of absolute ethanol. Fill a 1-dm polarimeter tube with the solution. Set the polarimeter at a zero reading by first viewing the sodium light through a 1-dm tube filled with absolute ethanol. Replace the tube containing the pure solvent with that containing the solution and determine both the direction and extent of the rotation of your sample. Use the average of three measurements as your observed rotation measurement. Calculate the specific rotation, $[\alpha]_D^{20}$, from either of the following relationships, which are exactly equivalent and differ only in the units of the concentration term c. The form on the left may be used for pure liquids or solutions, whereas the form on the right is used only for solutions. However, in the chemical literature, specific rotations are reported as if they were calculated using the equation on the right, that is, with the concentration c in g/100 mL.

$$[\alpha]_\lambda^t = \frac{\alpha}{c \times 1} \quad \text{or} \quad [\alpha]_\lambda^t = \frac{100 \times \alpha}{c \times 1}$$

where

α = the observed rotation
t = the temperature at which the measurement was made
λ = the wavelength of light used. This normally is the D line in the sodium spectrum, a doublet that appears at a wavelength of 5890–5896Å
c = concentration of solution in g/mL (left equation) or g/100 mL (right equation)
l = length of light path in dm

The optical rotation of (R)-(+)-limonene is reported as $[\alpha]_D^{20} = +125.6°$ (undil). For a typical commercial preparation (97% purity) the rotation is reported to be $[\alpha]^{24} = +106°$ ($c = 1$, CH_3OH). Record on the report form the specific rotation of your sample of limonene using the latter format. Return any unused limonene to your instructor in a properly labeled vial.

Note 1   Methods of heating are discussed in the General Instructions beginning on page 24 and in Experiment 3 beginning on page 53.

Note 2   Use a sharp knife to remove only the outer portion of the orange peel. Do not use the heavy white pulp that constitutes the bulk of an orange peel.

Note 3   If, at this stage of instruction, you have not yet reached a level of attainment necessary to understand instrumental methods of analysis, your instructor may wish to do tests 1, 3, and 4 as a demonstration project for the entire class. If a good commercial polarimeter is not available, a simple polarimeter can be assembled from inexpensive parts: Shavitz, R., "An Easily Constructed Student Polarimeter," *Journal of Chemical Education*, **1978,** *55,* 682.

REPORT 9

Name ____________________ Section ____________ Date ____________

Interpretation of the ir spectrum of limonene:

Type of bond	Wavenumber (cm^{-1}) and intensity[1]	Wavelength (microns, μ)	Remarks

Results of test with Br_2–CH_2Cl_2 solution:

Results of GLC study:

Results of polarimetery test: $[\alpha]_D$ = ________ c = ________

Yield of (+)-limonene: ________

[1]s = strong; m = medium; w = weak; b = broad

1. To what general class of natural products does limonene belong?

2. Draw five other structural isomers that possess the limonene skeleton. Name them.

3. Draw structures for the products that could result if limonene were treated with ozone and the resulting ozonide subjected to reductive hydrolysis.

4. The following natural products all may be steam distilled. Which would distill most readily? Why?

Product	*Boiling Point*
Eugenol	250°
Vanillin	285°
Citronellal	208°

5. Name essences or flavoring agents other than those listed in Question 4 that could probably be isolated by a steam distillation procedure.

6. If your unknown in Experiment 5 had consisted of benzoic acid and *p*-dichlorobenzene, could you have separated your mixture by steam distillation? Why?

7. What properties must an organic compound possess in order to be steam distilled?

8. Would it be correct to say that any organic compound possessing a pronounced aroma is distillable with steam? Explain.

9. It is good practice to take measurements of specific rotation on newly discovered optically active compounds at two substantially different concentrations, as your class was instructed to do. Show why this might be important by suggesting how you would determine whether an observed rotation of 180° is +180° or −180°. (Note that −180° and +180° come at the same point on the polarimeter dial.)

10. We have labeled the limonene obtained from orange peel as the (*R*)-enantiomer. Draw a broken line-wedge structural formula for limonene that shows this configuration.

Molecular Structures: Isomerism

The ability to visualize molecules in space is almost essential to understanding organic chemical behavior. Therefore, the objective of this exercise is to provide you with experience in building molecular models and drawing three-dimensional representations of organic structures. The building of molecular models will give you a feel for organic structures, will help you visualize the various spatial arrangements possible when carbon atoms are bonded to other groups or to other carbons, and will help you better understand the concept of isomerism.

PROCEDURE Use a model kit to construct a molecular model for each of the formulas or structures given in column A and make a three-dimensional drawing of your model using the solid line-broken line-wedge bond convention. The numbers in parentheses before some formulas indicate that more than one structure may be drawn for the formula. However, you need construct only one conformer or one isomer or one enantiomer for any molecular formula. All students will construct the same models for compounds (a), (b), (c), (d), (e), and for those molecules for which a structure rather than a molecular formula is given. Two of the compounds possible for formula (j) are drawn out for you as examples. Do not forget to draw your own example for formula (j). More specific instructions and possible variations that should be helpful in building your models are given in column B of Part I. Submit the completed exercise as your laboratory report.

PART I Structures from Molecular Formulas

A	B	C	D
Structural formula	Structural variations	Three-dimensional drawing	Name
(a) CH_3Cl	none		
(b) CH_4O	none		

A	B	C	D
Structural formula	Structural variations	Three-dimensional drawing	Name
(c) C_2H_6	*		
(d) C_2H_5Cl	*		
(e) C_3H_8	*		
(f-2) C_3H_7Cl	§		
(g-2) C_4H_{10}	*, §		
(h-5) C_4H_9Cl	§, †		
(i-9) $C_4H_8Cl_2$	*, §, †		
(j-15) C_4H_8O	§, †, ¶, +		(R)-3-Buten-2-ol
			(Z)-2-Buten-1-ol

A	B	C	D
Structural formula	Structural variations	Three-dimensional drawing	Name
(k) C_6H_{12}	+		
(l-10) $C_6H_{11}Cl$	+		
(m-4) $C_6H_{10}Cl_2$	*, †, +		
(n-6) C_4H_8	§, †		
(o-7) $C_4H_{10}O$	*, §, +		

* May be shown as Newman structure
§ Structural isomers
† Stereoisomers
+ Cyclic
¶ Functional group isomers

PART II   Structures from Structures

A	B	C
Structural formula	Three-dimensional drawing	Name

(p)
```
       CH=O
        |
 HO—C—H
        |
  H—C—OH
        |
      CH₂OH
```

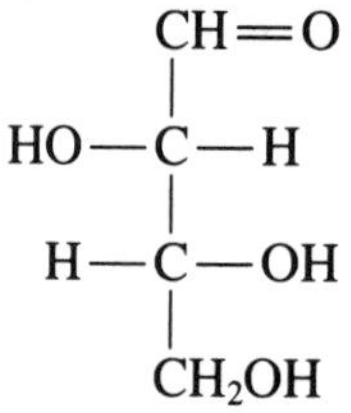

(q)

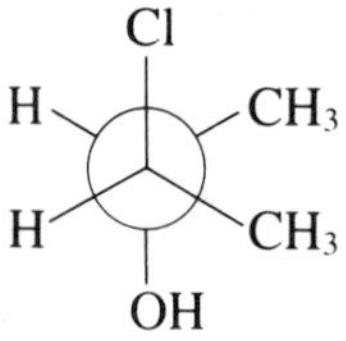

(r)

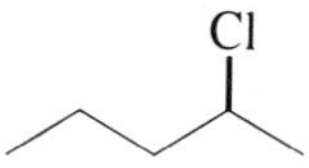

(s)

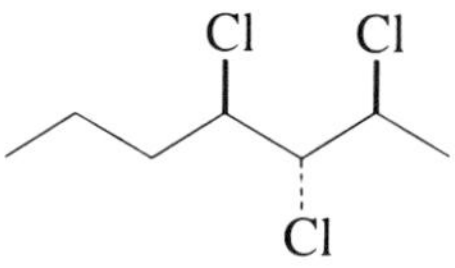

(t)

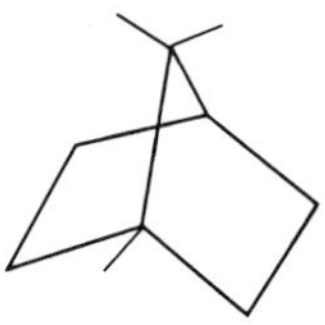

A Study of Reaction Rates— The Hydrolysis of tert-*Butyl* Chloride

A large number of organic reactions are those in which one group (a nucleophile, a Lewis base) displaces another. The group displaced is usually a weaker nucleophile and often is referred to as a good leaving group. Such reactions, called substitution reactions, have been studied in great detail with special emphasis upon the factors affecting the rate at which substitutions occur. Important rate-determining factors are: (1) the concentration of the reactants involved in the substitution, (2) the temperature at which the molecules react, (3) the nature of the leaving group and that of its replacement, and (4) the nature of the solvent in which the reaction takes place.

A knowledge of the variables that affect reaction rates is important to organic chemists, for it enables them to carry out reactions under optimum conditions for the preparation of a product in the greatest possible yield.

The hydrolysis of a tertiary alkyl halide is a chemical reaction that lends itself especially well to a rate study experiment. It is a classic example of a nucleophilic and unimolecular substitution reaction that is commonly designated as the S_N1 reaction. This reaction takes place in two principal steps:

Ionization step (slow step)

$$CH_3-\underset{\underset{CH_3}{|}}{\overset{\overset{CH_3}{|}}{C}}-Cl \longrightarrow CH_3-\underset{\underset{CH_3}{|}}{\overset{\overset{CH_3}{|}}{C^+}} + Cl^- \qquad (1)$$

Hydrolysis step (fast step)

$$CH_3-\underset{\underset{CH_3}{|}}{\overset{\overset{CH_3}{|}}{C^+}} + H_2O \longrightarrow CH_3-\underset{\underset{CH_3}{|}}{\overset{\overset{CH_3}{|}}{C}}-OH + H^+ \qquad (2)$$

Combining equations (1) and (2) gives the net equation

$$CH_3-\underset{\underset{CH_3}{|}}{\overset{\overset{CH_3}{|}}{C}}-Cl + H_2O \longrightarrow CH_3-\underset{\underset{CH_3}{|}}{\overset{\overset{CH_3}{|}}{C}}-OH + HCl \qquad (3)$$

137

Inasmuch as no reaction sequence can proceed faster than the slowest step, Equation (1) is the **rate-determining step** and the speed with which the reaction takes place is dependent upon the concentration of the *tert*-alkyl halide alone. A reaction of this type is said to follow first-order kinetics and the reaction rate may be expressed as

$$\text{rate} = k[RX]$$

where k is the reaction rate constant expressed in moles/liter $\times$ sec^{-1}, and $[RX]$ is the concentration of the alkyl halide. The object of the present experiment is to determine a value for k. As may be seen from Equation (3), one mole of HCl is produced for each mole of alkyl halide undergoing hydrolysis. By measuring the time required for a predetermined concentration of alkyl halide to undergo hydrolysis, measured by an indicator change when the HCl produced is neutralized by a known volume of standardized base, you will be able to calculate a rate constant for the reaction from the expression

$$kt = 2.303 \log\left(\frac{c}{c - x}\right)$$

where c is the initial concentration of alkyl halide and $(c - x)$ is the concentration after x moles have reacted during time t. The value of k is constant only when the reaction is carried out in the same solvent system and at the same temperature. For this reason it is recommended that burets be filled from stock bottles that have been allowed to equilibrate with room temperature. Also, it is important that reaction times be carefully measured. This is best accomplished by allowing students to work in pairs during this experiment—one mixing reagents, the other carefully noting and recording the time when reaction is begun and completed, along with other data.

PROCEDURE

Caution! *Certain of the alkyl halides are cancer suspect. Although* tert-*butyl chloride is not on the cancer suspect list at present,* tert-*butyl bromide is. Therefore, use a pipet aid, not your mouth, to draw up the* tert-*butyl chloride. Avoid contact with this or any alkyl halide, and avoid breathing the vapor.*

Prepare a 0.1M solution of *tert*-butyl chloride in acetone by adding 0.55 mL of *tert*-butyl chloride to a 125-mL Erlenmeyer flask containing 50 mL of *dry*, reagent-grade acetone. Use a 1-mL graduated pipet to accurately measure the alkyl halide. Transfer a portion of the alkyl halide-acetone solution to a 10-mL buret and clamp the buret in place. Fill a 25-mL buret with distilled water. A buret assembly having a white ceramic base is advantageous for this experiment.

Pipet 0.5 mL of 0.1M sodium hydroxide solution into a clean, dry 50-mL Erlenmeyer flask, add 14.5 mL of distilled water from the buret, and add 3 drops of Bromphenol Blue indicator. Add 5mL of *tert*-butyl chloride-acetone solution from the first buret to the sodium hydroxide solution as quickly as possible. Swirl the flask while making the addition to ensure good mixing, and start timing the reaction with a stop watch or wrist watch when one-half of the alkyl halide has been added. Record on the report form the total time required for the indicator to change from blue to yellow.

Discard the reaction mixture, rinse your flask with distilled water and then with a little acetone. Allow the acetone to completely evaporate before making a second determination, using the same quantities as before. Again, note and record the reaction time. The average of the two time values will be taken as *t* for one-tenth of the *tert*-butyl chloride to undergo hydrolysis.

In the second part of the experiment you will study what effect, if any, a change in the concentration of sodium hydroxide will have upon the rate of hydrolysis of *tert*-butyl chloride. Repeat the previous procedure but this time add 1 mL of $0.1M$ sodium hydroxide to a 50-mL Erlenmeyer flask containing 14 mL of distilled water and 3 drops of Bromphenol Blue indicator. Add the same volume of *tert*-butyl chloride-acetone as before. Again, note and record the time required for an indicator color change. Repeat and use the average of time values as *t* for one-fifth of the *tert*-butyl chloride to undergo hydrolysis.

In the third part of the experiment you will study what effect a change in concentration of alkyl halide will have upon the rate of hydrolysis. Prepare a sodium hydroxide solution by adding 1 mL of $0.1M$ sodium hydroxide to 14 mL of distilled water from the buret. Add 2.5 mL of dry acetone to the basic solution and again 3 drops of indicator. Add 2.5 mL of *tert*-butyl chloride-acetone solution from the first buret. Note and record the time required for an indicator color change. Repeat and use the average time values as *t* for two-fifths of the alkyl halide to undergo hydrolysis.

Record all concentration and time values on the report form. From these values, calculate a value for the reaction rate constant, *k*.

Name Section Date

The Hydrolysis of *tert*-Butyl Chloride

Part	Vol. NaOH	Vol. RX	[OH$^-$]	[RX]	t(sec)	$\dfrac{c}{c-x}$	k^1
I Trial 1 2							
II Trial 1 2							
III Trial 1 2							

Sample calculation (for Part I):

$$kt = 2.303 \log \frac{[RX]}{[RX] - [OH^-]}$$

$$= 2.303 \log \frac{0.025}{0.025 - 0.0025}$$

$$= 2.303 \log 10/9 = 2.303 \log 1.11$$

$$= 0.1043$$

$$k = \frac{0.1043}{t} \frac{\text{moles sec}^{-1}}{\text{liter}}$$

---

[1]An average value for k is $3.54 \times 10^{-3} \dfrac{\text{moles sec}^{-1}}{\text{liter}}$ for the solvolysis of *tert*-butyl chloride.

1. What conclusion may be drawn regarding the value of k as a function of concentration of either $[OH^-]$ or $[RX]$?

2. The solvent in which the hydrolysis of *tert*-butyl chloride took place was approximately 75% water–25% acetone by volume. What effect would increasing the percentage of acetone have upon the rate of hydrolysis? What effect upon the rate of hydrolysis could be expected had you used isopropyl alcohol instead of acetone as the organic component of our solvent? Explain.

3. The intermediate in the S_N1 reaction route is a carbocation. What options other than combination with a nucleophile are open to a carbocation?

4. If the alkyl halide undergoing hydrolysis via an S_N1 mechanism were an optically pure compound, would the product be optically active? Explain.

5. Experimental evidence shows that complete racemization does not occur when an optically active compound undergoes an S_N1 displacement. Why are not equal amounts of each isomer produced?

6. List the following halogen compounds in order of reactivity toward S_N1 displacement: (1) *tert*-butyl chloride, (2) allyl chloride, (3) isobutyl chloride, (4) benzyl chloride, (5) *sec*-butyl chloride; (6) neopentyl chloride.

Reactions of the Alcohols and Phenols

The alcohols may be represented by the general formula R—OH. The reactions of the alcohols, if R is a saturated alkyl group, will be mainly those of the relatively reactive hydroxyl group rather than those of the comparatively inert alkyl group.

The phenols, unlike the alcohols, have the hydroxyl group bonded to a carbon atom that forms part of the aromatic ring.

This feature greatly alters the properties of the phenolic hydroxyl and serves to distinguish it from hydroxyl groups of the alcohols.

The following tests and reactions illustrate some of the general properties of the alcohols and phenols.

Precautions.   *This experiment requires the use of various alcohols and phenols, sodium metal, sodium hydroxide, hydrochloric acid, bromine, and potassium dichromate. See HAZARD CATEGORIES 1, 2, 4, and 6, page 4.*

PART I   Properties and Reactions of Alcohols

PROCEDURE   **A.   Solubility in Water and in Acidic Solutions**

Into four separate test tubes place small samples (about 10 drops each) of the following: *n*-butyl alcohol, *tert*-butyl alcohol, cyclohexanol, and benzyl alcohol.

Add 5 mL of water to each sample, mix, and observe. Next, place all test tubes in a water bath and warm to 65°. Test a drop of each solution on red litmus paper. Remove all samples from the water bath and allow them to cool to room temperature. Observe and record your results.

Repeat the above but instead of water, use equivalent volumes of 85% phosphoric acid. Mix by shaking each solution as before but do not heat. Again, observe and record your results.

B. Reaction with Sodium

Water reacts with metallic sodium to form hydrogen and sodium hydroxide. In a similar manner alcohols react with sodium to form hydrogen and sodium alkoxides.

$$2\ R\!-\!OH + 2\ Na \longrightarrow 2\ RO^-Na^+ + H_2$$

Phenols react with sodium to yield hydrogen and sodium phenoxide.

$$2\ Na + 2 \underset{\text{Phenol}}{\overset{\text{OH}}{\bigcirc}} \longrightarrow 2 \underset{\substack{\text{Sodium} \\ \text{phenoxide}}}{\overset{\text{O}^-\text{Na}^+}{\bigcirc}} + H_2$$

Place 2 mL of each of the following alcohols in separate, *dry* test tubes: (1) methyl alcohol, (2) isopropyl alcohol, (3) *n*-propyl alcohol. Add a small piece of sodium metal (about the size of a BB shot) to each test tube and observe the result. Allow the reaction to proceed to completion. Record your results on the report form.

C. Reaction of Alcohols with Lucas Reagent

The Lucas test for distinguishing between primary, secondary, and tertiary alcohols (the class of an alcohol is the same as the class of its alkyl group) is based on the relative rates of reaction of the different classes with hydrogen chloride. The Lucas reagent is a solution of zinc chloride in concentrated hydrochloric acid.[1] A tertiary alcohol reacts rapidly with the reagent to give an insoluble alkyl chloride that appears as a cloudy dispersion or as a separate layer in the solution. A secondary alcohol gives a clear solution at first (as the alcohol dissolves), which becomes cloudy *within five minutes*. A primary alcohol dissolves to produce a solution that remains clear for several hours.

$$R\!-\!OH + HCl \xrightarrow{\ \text{ZnCl}_2\ } R\!-\!Cl + H_2O$$

Add 1 mL of the Lucas reagent to each of three test tubes. Add 4–5 drops of the alcohol to be tested, mix well, and record the length of time it takes for the

[1]See Instructor's Guide.

mixture to become cloudy or to separate into two layers. Carry out the tests with: (1) *n*-butyl alcohol, (2) *sec*-butyl alcohol, and (3) *tert*-butyl alcohol. Record your results on the report form.

D. Oxidation of Alcohols

When the saturated, aliphatic alcohols are oxidized under relatively mild conditions, the product of the reaction depends upon the class of the alcohol being oxidized. Thus, primary alcohols give first aldehydes and finally carboxylic acids as oxidation products.

$$R-CH_2-OH \xrightarrow{[O]} R-C\!\!\overset{\displaystyle O}{\underset{\displaystyle H}{<}} \xrightarrow{[O]} R-C\!\!\overset{\displaystyle O}{\underset{\displaystyle OH}{<}}$$

| Primary alcohol | Aldehyde | Carboxylic acid |

By proper choice of the oxidizing agent it is often possible to stop the reaction at either the aldehyde or carboxylic acid stage. Secondary alcohols give ketones as oxidation products.

$$\overset{\displaystyle OH}{\underset{\displaystyle |}{R-CH-R'}} \xrightarrow{[O]} \overset{\displaystyle O}{\underset{\displaystyle ||}{R-C-R'}}$$

Secondary alcohol   Ketone

Tertiary alcohols are inert toward most mild oxidizing agents.

$$\overset{\displaystyle OH}{\underset{\displaystyle \underset{\displaystyle R''}{|}}{R-C-R'}} \xrightarrow{[O]} \text{No reaction}$$

Tertiary alcohol

Although hexavalent chromium compounds are known to be carcinogenic, at present there are no satisfactory substitutes for the hexavalent chromium oxidizing agents. Therefore, we must concentrate our efforts on learning to use them safely. Fortunately, few of them are volatile; therefore, the principal hazard is contact with the body. Also, the trivalent chromium compounds are not thought to be carcinogenic, so reduction of the hexavalent species is a relatively simple route to destruction and disposal of these materials.

Place 2 mL of each of the following alcohols in separate, clean 50-mL Erlenmeyer flasks: (1) ethanol, (2) isopropyl alcohol, and (3) *tert*-butyl alcohol. To each alcohol add 5 mL of an oxidizing solution prepared from 5 g of potassium dichromate in 50 mL of water and 5 mL of concentrated sulfuric acid. Mix the reagents by swirling the flasks and warm on the steam bath or hot plate. Observe any changes. Remove the flask containing the *tert*-butyl alcohol and set it aside. Bring the mixtures containing ethanol and isopropyl alcohol to a boil, either on the hot plate or on a wire gauze over a burner. If you use a burner, extinguish it; if you use a hot plate, turn it off. Using tweezers, place over the mouth of each flask a strip of moistened blue litmus paper. After 10 minutes, inspect each flask and record your observations. Write the equations for the reactions (if any) that took place.

Cleanup: Your instructor will inform you of the proper methods of disposal of the chromium-containing residues from your experiment.

E.   The Iodoform Reaction

Alcohols that have the structure $\overset{\overset{\text{OH}}{|}}{\text{CH}_3\text{—CH—R}}$ (R==H, alkyl, or aryl) and al-

dehydes or ketones that have the structure $\overset{\overset{\text{O}}{\|}}{\text{CH}_3\text{—C—R}}$ (R==H, alkyl, or aryl) react with an alkaline solution of the halogens (hypohalite solutions) to yield the corresponding haloforms. Indeed, the **haloform reaction** is often performed as a diagnostic test on an unknown compound suspected of having the above structural arrangements. The preparation of iodoform may be illustrated by the following reaction sequence.

$$2 \text{ NaOH} + \text{I}_2 \longrightarrow \text{NaI} + \text{NaOI}$$

$$\overset{\overset{\text{OH}}{|}}{\text{R—CH—CH}_3} + \text{NaOI} \longrightarrow \overset{\overset{\text{O}}{\|}}{\text{R—C—CH}_3} + \text{NaI} + \text{H}_2\text{O}$$

$$\overset{\overset{\text{O}}{\|}}{\text{R—C—CH}_3} + 3 \text{ NaOI} \longrightarrow \overset{\overset{\text{O}}{\|}}{\text{R—C—CI}_3} + 3 \text{ NaOH}$$

$$\overset{\overset{\text{O}}{\|}}{\text{R—C—CI}_3} + \text{NaOH} \longrightarrow \overset{\overset{\text{O}}{\|}}{\text{R—C—O}^-\text{Na}^+} + \underset{\text{Iodoform}}{\text{CHI}_3}$$

Chloroform and bromoform may be prepared in a similar manner, but, inasmuch as they are liquids, they are less useful for diagnostic purposes. Iodoform, on the other hand, is a yellow solid (mp 119–120°) with a sharp, easily recognizable odor.

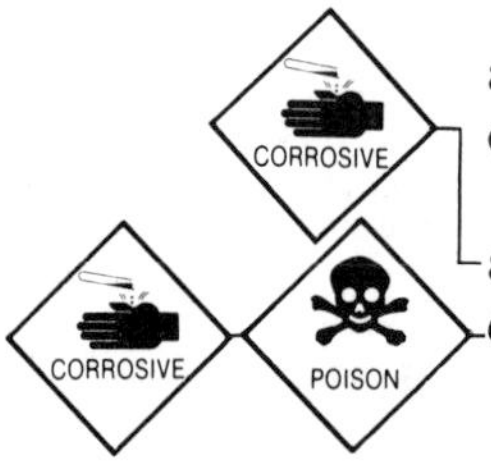

To 0.5 mL of acetone (about 10 small drops) in a test tube add 5 mL of water and 1 mL of dilute (6*N*) sodium hydroxide. To the alkaline solution of acetone add, dropwise, an iodine-potassium iodide solution.[1] Continue to add iodine until the

---

[1]See Instructor's Guide.

brown color no longer is discharged and only a light yellow color remains. If less than 1 mL of the iodine-potassium iodide solution is used, heat the test tube in the water bath at 60° and add any additional iodine-potassium iodide needed. Collect the precipitate that forms on the Hirsch funnel, wash it on the filter with 5 mL of (5%) dilute sodium hydroxide, and dry on a piece of clay plate or on a large piece of filter paper placed on a watch glass. Determine the melting point of the precipitate.

Repeat the iodoform test using (a) ethyl alcohol, (b) methyl alcohol, (c) iso-propyl alcohol, and (d) *n*-propyl alcohol. (Note 1) You may omit the melting point determinations. Record your results on the report form.

Note 1   For alcohols having very low water solubility it may be necessary to dissolve the sample (10 drops or 0.1 g) in 5 mL of dioxane. In such instances it is advisable to run a blank on the dioxane, for some samples of dioxane have given positive iodoform tests.

F.   Determination of an Unknown

From your instructor obtain an unknown alcohol and determine its identity by a series of tests that will include one or more of the following: the Lucas test, oxidation with dichromate-sulfuric acid reagent, or the haloform reaction. Your unknown will be one of those listed in Table 12.1. (**Caution!   Instructor: Read Note 1, p. 57.**)

Table 12.1   Structures and Boiling Points (°C) of Some Alcohols

Compound	Formula	Boiling point
2-Propanol	CH_3—CH—CH_3 with OH on CH	82
2-Methyl-2-propanol	CH_3—C—CH_3 with CH_3 above and OH below central C	83
1-Propanol	$CH_3CH_2CH_2OH$	98
2-Butanol	CH_3CH_2—CH—CH_3 with OH below CH	100
2-Methyl-2-butanol	CH_3CH_2—C—CH_3 with CH_3 above and OH below central C	102

Table 12.1   *(Continued)*

Compound	Formula	Boiling point
2-Methyl-1-propanol	CH_3 $\vert$ CH_3—CH—CH_2OH	108
3-Pentanol	CH_3CH_2—CH—CH_2CH_3 $\vert$ OH	116
1-Butanol	$CH_3CH_2CH_2CH_2OH$	118
2-Pentanol	OH $\vert$ CH_3—CH—$CH_2CH_2CH_3$	119
3-Methyl-1-butanol	CH_3 $\vert$ CH_3—CH—CH_2CH_2OH	130
4-Methyl-2-pentanol	CH_3 OH $\vert$ $\vert$ CH_3—CH—CH_2—CH—CH_3	132
Cyclopentanol	CH_2—CH_2 $\vert$ $\diagdown$ $\qquad\qquad$ CH—OH $\vert$ $\diagup$ CH_2—CH_2	141
2-Methyl-2-hexanol	CH_3 $\vert$ $CH_3(CH_2)_3C$—CH_3 $\vert$ OH	143

A boiling point determination is a part of this exercise and should be performed as the first step. This is best done by a small-scale distillation (Experiment 3) on a sample of 5 mL or more, but, if your sample is but a milliliter or two, it may be done as follows. (*Caution!* *Instructor: Read Note 1, p. 57.*)

Construct a small boiling point tube, which resembles a small test tube without a lip, by sealing one end of an 8- to 10-cm length of 4- to 6-mm glass tubing. Attach the boiling point tube to a thermometer with a small rubber band or a slice of rubber tubing exactly as described in Experiment 1 (page 36) for the attachment of melting point tubes. The bottom of the boiling point tube should be opposite the bottom of the thermometer, and the rubber band should be placed near the top of the boiling point tube. Next, construct a capillary ebullition tube by sealing a melting point capillary about 1 cm from an *open end*. This is best done with longer than normal melting point tubes that are open at both ends, using a small glass blowing torch to make a seal far enough from one end to allow you to handle the tube conveniently and then cutting the tube cleanly on the shorter end about 1 cm below the seal. The best ebullition tubes are about 1 cm longer than the boiling point tube for easy insertion and removal. Your instructor can provide both the boiling point tube and the capillary ebullition tubes. (Note 1)

Slide the capillary ebullition tube carefully into the boiling point tube to avoid breakage. Using a pipet or syringe, add to the boiling point tube the liquid whose boiling point is to be determined until the level of liquid in the tube is 2–3 mm above the seal in the ebullition tube. Mount the thermometer and attached micro boiling point tube (Fig. 12.1) in either a melting point bath (Fig. 1.1) or in a Thiele melting point apparatus (Fig. 1.2), making sure that the level of the oil is about 2 cm below the rubber band. Turn on the heat source and heat the bath as if determining a melting point, except that the bath should be heated fairly quickly up to the point where a *rapid and continuous* stream of bubbles comes out of the capillary ebullition tube. Turn off the heat source and allow the bath to cool *slowly* while carefully observing the capillary ebullition tube. The rate of bubbling will decrease, stop, and then the liquid will begin to rise in the capillary ebullition tube, finally filling the tube. Record on your report form the temperatures at which (a) the liquid first starts to enter the capillary ebullition tube and (b) the tube is just filled with liquid. These temperatures

Figure 12.1   Micro boiling point tube.

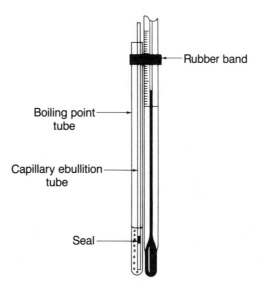

define the micro boiling point range. If you wish to repeat the determination, you will have to remove the thermometer, allow the attached tubes to cool, remove the capillary ebullition tube, shake out the liquid in the capillary, and replace the capillary ebullition tube in the boiling point tube.

PART II   Properties and Reactions of Phenols

PROCEDURE   **A.   Solubility in Water and Alkaline Solutions**

When the phenyl group is present as a substituent in an aliphatic compound (alcohol, aldehyde, ketone, acid, etc.), its effect on the solubility of the compound in water is approximately equivalent to that of a four-carbon alkyl group. Thus, benzyl alcohol has about the same solubility in water as *n*-amyl alcohol. This effect carries over to some extent to aromatic compounds as well. Benzoic acid is only slightly less soluble than *n*-valeric acid and phenol has approximately the same solubility as *n*-butyl alcohol.

Phenols are more acidic than the alcohols but less so than the carboxylic acids or carbonic acid. Phenols thus will form salts when treated with sodium hydroxide but not with sodium carbonate. This difference in acid strength serves to distinguish phenols from alcohols and from carboxylic acids.

	OH		O⁻Na⁺	

$$\text{NaOH} + \underset{\text{Phenol}}{\overset{\text{OH}}{\bigcirc}} \longrightarrow \underset{\substack{\text{Sodium} \\ \text{phenoxide}}}{\overset{\text{O}^-\text{Na}^+}{\bigcirc}} + \text{H}_2\text{O}$$

Into three separate test tubes, place small samples—either 10 drops of liquid or a few small crystals (0.1 g) of a solid—of the following: phenol, *p*-chlorophenol, and 2-naphthol. Add 5 mL of water to each sample, mix, and observe. Next, place all test tubes in a water bath and warm to 65°. Test a drop of each solution on blue litmus paper. Remove all samples from the water bath and allow them to cool to room temperature. Observe and record your results. Save all solutions for Procedure C.

Repeat Procedure A but, instead of water, use equivalent volumes of 5% sodium hydroxide solution. Mix by shaking each solution as before but do not heat. Again, observe and record your results.

Repeat Procedure A but, instead of water, use equivalent volumes of 5% sodium bicarbonate solution. Mix by shaking each solution as before but do not heat. Again, observe and record your results.

B.   Reaction with Bromine

Bromine reacts very readily with phenols to introduce one or more bromine atoms in positions *ortho* or *para* to the phenolic hydroxyl group.

$$\text{OH} + 3\ Br_2 \longrightarrow \text{(2,4,6-tribromophenol)} + 3\ HBr$$

Add bromine water dropwise to each of the solutions you prepared in Procedure A until the color of bromine is no longer discharged. Shake well after each addition. Record your results.

C.   Ferric Chloride Test

Phenols and enols often react with ferric chloride to produce characteristic colors. Not all phenols and enols give colors; therefore, a negative test is not necessarily significant.

To very dilute (0.1%) solutions of (1) phenol, (2) resorcinol, and (3) salicylic acid made by dissolving a few crystals of each in 5 mL of water, add a drop of 1% ferric chloride solution. Try the same tests on a sample of ethyl alcohol and also on a methanolic solution of ethyl acetoacetate. Record your results on the report form.

Note 1   Ebullition tubes may be made fairly easily by joining two capillary melting point tubes at their sealed ends in the small, hot flame of an air-gas glass-blowing torch. Unsealed melting point tubes may be joined in the same way. After the joint has cooled, the tube is cut about 1 cm above the seal on one end. Boiling point tubes can be made from clean Pasteur pipets (salvaged from those with broken tips), although the diameter of the pipets is usually slightly larger than optimal.

Name   *Section*   *Date*

Properties and Reactions of the Alcohols

Solubility in Water and in Phosphoric Acid

Alcohol	Structure	Cold water	Hot water	85% H_3PO_4
n-Butyl alcohol	$CH_3(CH_2)_3OH$			
tert-Butyl alcohol	$CH_3-\overset{\overset{\displaystyle CH_3}{\displaystyle \vert}}{\underset{\underset{\displaystyle CH_3}{\displaystyle \vert}}{C}}-OH$			
Cyclohexanol	⬡—OH			
Benzyl alcohol	⬡—CH_2—OH			

Reaction with Sodium

Alcohol	Structure	Result
Methyl alcohol	CH_3-OH	
Isopropyl alcohol	$CH_3-\overset{\underset{\underset{\displaystyle CH_3}{\displaystyle \vert}}{}}{CH}-OH$	
n-Propyl alcohol	$CH_3-CH_2-CH_2-OH$	

Reaction with Lucas Reagent

Alcohol	Time for reaction	Description of reaction mixture
n-Butyl alcohol		
sec-Butyl alcohol		
tert-Butyl alcohol		

Oxidation of Alcohols

Alcohol	Structure	Color changes observed	Color of litmus paper
Ethanol	CH_3-CH_2-OH		
Isopropyl alcohol	$CH_3-CH-OH$ $\quad\quad\; \mid$ $\quad\quad CH_3$		
tert-Butyl alcohol	$\quad\quad CH_3$ $\quad\quad\; \mid$ CH_3-C-OH $\quad\quad\; \mid$ $\quad\quad CH_3$		

Write balanced equations for the reactions (if any) that took place.

Ethanol:

Isopropyl alcohol:

Name _____   Section _____   Date _____

tert-Butyl alcohol:

Results of the Iodoform Reaction

Compound	Structure	Result
Acetone	$CH_3 - \overset{\overset{\displaystyle O}{\|\|}}{C} - CH_3$	
Ethyl alcohol	$CH_3 - CH_2 - OH$	
Methyl alcohol	$CH_3 - OH$	
Isopropyl alcohol	$CH_3 - \overset{\overset{\displaystyle OH}{\|}}{\underset{\underset{\displaystyle H}{\|}}{C}} - CH_3$	
n-Propyl alcohol	$CH_3 - CH_2 - CH_2 - OH$	

Melting Point of Iodoform

Experimental _____ °C

Literature _____ °C

Write balanced equations for the formation of iodoform from each compound above that gave a positive test.

Determination of an Unknown

Unknown no. _____ ;   Boiling point _____ °C

Possible compounds _____

_____

Results of distinguishing tests:

Lucas test _____

Iodoform reaction _____

Oxidation reaction _____

Identity of unknown _____

Properties and Reactions of Phenols

Solubility in Water, 5% Sodium Hydroxide, and 5% Sodium Bicarbonate

Compound	Structure	Cold water	Hot water	5% NaOH	5% NaHCO$_3$
Phenol	OH (benzene ring)				
p-Chlorophenol	Cl—(benzene ring)—OH				
2-Naphthol	(naphthalene ring)—OH				

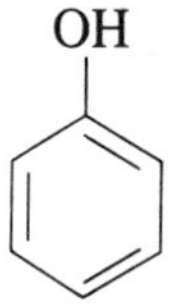

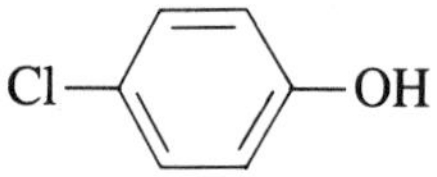

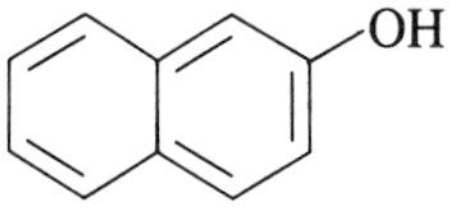

Reaction with Bromine

Compound	Result with bromine water
Phenol	
p-Chlorophenol	
2-Naphthol	

Name   Section   Date

Reaction with Ferric Chloride

Compound	Structure	Result
Phenol		
Resorcinol		
Salicylic acid (consult text)		
Ethyl alcohol		
Ethyl acetoacetate (consult text)		

1. Can you offer an explanation for the variation in reaction rate observed when primary, secondary, and tertiary alcohols react with the Lucas reagent?

2. An unknown alcohol gives a negative Lucas test and a positive iodoform test. Suggest a possible structure for the alcohol.

3. Suggest a simple method for the separation of phenol from a solution that contains both phenol and cyclohexanol.

4. An unknown reacts with sodium to form a sodium salt but further tests show the compound to be an aliphatic hydrocarbon. What structural unit is probably present in the hydrocarbon?

5. Theoretically, how much iodoform could be prepared from 25 mL of 95% ethanol (sp. gr. 0.79)?

6. What is a medicinal use for iodoform?

7. Beginning with benzene, show how you might prepare benzoic acid, C_6H_5-COOH, using the iodoform reaction in one step of the synthesis.

8. Only one aldehyde and one primary alcohol give a positive iodoform reaction. What compounds are these? Explain why they alone of their class give a positive reaction.

9. Systematic names are assigned to the compounds in Table 12.1. Can you assign another acceptable name to each?

10. Why do alcohols that show little tendency to dissolve in water appear to dissolve readily in 85% phosphoric acid? Would you describe the latter behavior as solubility or a chemical reaction? Explain.

11. Explain why ethyl acetoacetate was expected to react with ferric chloride in Part II-C of this experiment. (*Hint:* Review the sections describing enolates and "tautomerism" in your text.)

12. Draw the structures of 2-naphthol and benzhydrol (Experiment 13). Describe a simple chemical test for distinguishing between these two compounds.

13. Explain the experimental observations made during a micro boiling point determination: the initial, somewhat erratic bubbling; the steady bubbling until the heat source is removed; the decrease in bubbling; and, finally, the capillary ebullition tube filling with liquid. Why is the boiling range defined as stated on page 149?

*14. Both ir and nmr spectra of alcohols and phenols show similar absorption ranges for the hydroxyl group. In which regions of both types of spectra must we seek additional absorption evidence that will help us to distinguish a phenol from an aliphatic alcohol?

*15. The aliphatic alcohols and ethers are isomeric; seven structures may be written for the molecular formula, $C_4H_{10}O$. If you had to identify an unknown of this formula using a spectroscopic technique, tell exactly how you would go about it.

*16. If the nmr spectrum of the compound in Question 15 consisted of only two signals—one a triplet at $\delta 1.15$ and the other a quadruplet at $\delta 3.48$—which of the seven compounds would be indicated?

Reactions of Aldehydes and Ketones

The aldehydes, $R-\overset{\overset{\displaystyle H}{|}}{C}=O$, and the ketones, $R-\overset{\overset{\displaystyle R}{|}}{C}=O$, both have the carbonyl group, $\overset{\displaystyle \diagdown}{\underset{\displaystyle \diagup}{C}}=O$, as a functional group and usually react with the same reagents to give similar products. In general, the principal differences in their reactions are in the relative rates of reaction. The aldehydes usually react more rapidly than do the ketones and give reactions that more nearly go to completion. Most of the reactions of both classes of compounds take place by addition across the carbon-oxygen double bond, frequently followed by water loss. This experiment illustrates several reactions of both aldehydes and ketones as useful chemical tests for distinguishing between the two classes and includes a simple synthetic procedure for the reduction of aldehydes and ketones.

> **Precautions.** *This experiment requires the use of aldehydes, ketones, ether, and methanol—all flammable liquids; hydrazine derivatives, sodium and ammonium hydroxides. See HAZARD CATEGORIES 1, 2, 3, and 4, page 4.*

PART I Chemical Tests

PROCEDURE **A. Test for the Carbonyl Group**

One of the simplest and most effective chemical tests for the carbonyl group of an aldehyde or ketone is based on the reaction of carbonyl compounds with 2,4-dinitrophenylhydrazine (**Caution!** *Flammable Solid.*) to form 2,4-dinitrophenylhydrazones. If an unknown compound reacts with 2,4-dinitrophenylhydrazine to form a yellow to red precipitate, the test is positive, and the compound probably contains a carbonyl group. This same reaction can be used to prepare derivatives of carbonyl compounds, as described in Experiment 14.

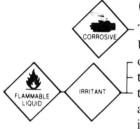

2,4-Dinitrophenylhydrazine   A 2,4-dinitrophenylhydrazone

There are now two methods for carrying out this test. If you are going to do Part B of this experiment, use the dimethylformamide method. Otherwise use the method for which your instructor provides the reagents.

(a) Sulfuric Acid—Ethanol Method

To each of four test tubes add 1 mL of 2,4-dinitrophenylhydrazine reagent. (Note 1) Unless your instructor asks you to test a different series of carbonyl compounds, obtain four clean Pasteur pipets, and to the first test tube add 2 drops of heptaldehyde, to the second add 2 drops of acetone, to the third add 2 drops of cyclohexanone, and to the fourth add 2 drops of acetophenone. Shake the test tubes to mix the reactants and allow them to stand for 15 minutes. The formation of a yellow to red precipitate is a positive test.

(b) Dimethylformamide Method

Caution! *Dimethylformamide is an irritant. Do not spill this material on yourself, and do not inhale its vapors. Dimethylformamide is very soluble in water and may be washed off the body with liberal use of soap and water.*

To each of four *labeled* test tubes or, preferably, 25-mL Erlenmeyer flasks add 0.5 mL of a 10% solution of 2,4-dinitrophenylhydrazine in dimethylformamide. (Note 2) Unless your instructor asks you to test a different series of carbonyl compounds, obtain four clean Pasteur pipets, and to the first flask add 3 drops of methyl *tert*-butyl ketone (pinacolone), to the second add 3 drops of acetone, and to the third and fourth add 3 drops of two carbonyl compounds chosen from the following (Note 3): diethyl ketone, heptaldehyde, cyclopentanone, or cyclohexanone. Swirl or shake the flasks to mix the reactants, then add 2 drops of concentrated hydrochloric acid to each flask. A precipitate may or may not form at any stage of this test. Allow the reaction mixtures to stand at room temperature for 15 minutes, then add 10 mL of 2M hydrochloric acid to each flask. The formation of a yellow to red precipitate that does not dissolve in the dilute hydrochloric acid is a positive test. Save the contents of the flasks for Part B.

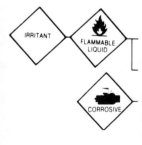

B.   TLC Separation of 2,4-Dinitrophenylhydrazones

In this part of the experiment you will determine whether it would be possible to separate the 2,4-dinitrophenylhydrazones prepared in Part A by thin layer chromatography. The procedure used will be very similar to that described in Experiment 7, Part III, which you may wish to review.

Assemble the components necessary for a TLC separation: a 400-mL beaker with a piece of aluminum foil for a cover, four micropipets, and a 50 × 100-mm TLC plate. Lightly draw a sampling line on the TLC plate as described in Experiment 7, Part III. Near the very bottom of the plate, gently write four identifying letters or numbers so that you will be able to identify the spots. To each of the four flasks (or

test tubes) from Part A2 add enough ordinary ether to form a layer about 1–1.5 cm deep and carry out the following procedure on each flask. Stopper the flask and shake well to extract the 2,4-dinitrophenylhydrazone into the ether layer, periodically removing the stopper to release the pressure. Add more ether if necessary to maintain at least a 1-cm layer. Allow the layers to separate. If you are using test tubes and the ether layer is not close enough to the mouth of the test tube to reach with a micropipet, transfer the ether layer with a Pasteur pipet to a very small Erlenmeyer flask or beaker. If you are using 25-mL Erlenmeyer flasks, a transfer should not be required. Dip a micropipet *into the ether layer* in each flask and place a 1- to 2-mm spot on the sampling line above the appropriate letter. After the spots dry, you may respot any that looks very weak, but this is usually not necessary. Pour into the

400-mL beaker 8 mL of hexane and 2 mL of ether, swirl the liquids quickly to mix them, place the TLC plate in the developing solvent (making certain that the spots are above the liquid level), and cap the beaker tightly with the aluminum foil. Allow the TLC plate to develop until the solvent front ascends to within 1.5–2.0 cm of the upper edge of the plate. Remove the aluminum foil and quickly mark the location of the solvent front (the solvent evaporates rapidly!). Remove the plate, allow it to dry, and determine the R*f* values of your 2,4-nitrophenylhydrazones. Sometimes there will be a faint row of four yellow spots, each having the same R*f* value, below the 2,4-dinitrophenylhydrazone spots caused by traces of 2,4-dinitrophenyl-hydrazine hydrochloride. Pour the reaction mixtures in the flask and the developing

solvent into the waste receptacle in the hood. The yellow residues may be removed from your glassware with acetone. (***Caution!*** *Acetone is very flammable.*)

C. Tests for Distinguishing between Aldehydes and Ketones

The chemical tests used to distinguish between aldehydes and ketones depend upon the difference in ease of oxidation of the two classes of compounds or upon differences in reactivity toward certain reagents. Although both aldehydes and ketones may be oxidized to acids, ketones are oxidized only under vigorous oxidizing conditions (that is, stronger reagents and high temperatures). Under these conditions cleavage of the carbon chain occurs. Aldehydes, on the other hand, are readily oxidized under very mild conditions without cleavage of the carbon chain. The principal oxidizing agents used in the test reactions are the silver ion and the cupric ion. These ions are the active agents in two test solutions called, respectively, Tollens' reagent and Benedict's solution. The cupric ion is also the active agent in Fehling's solution. Another useful test reaction is based on a difference in reactivity of the two classes toward a test solution called Schiff's reagent.

You will find that most of these tests give the best results with low molecular weight water-soluble aldehydes, such as formaldehyde, acetaldehyde[1], and propionaldehyde. However, Schiff's reagent is very sensitive and gives good positive tests with most aldehydes.

1. **Tollens' reagent** is an ammoniacal solution of silver oxide. The reagent is reduced by aldehydes to metallic silver, and the aldehyde is oxidized to the corresponding acid. Ketones do not react with the reagent.

$$2\ AgNO_3 + 2\ NaOH \longrightarrow \quad Ag_2O + H_2O + 2\ NaNO_3$$
$$Ag_2O + 4\ NH_4OH \longrightarrow \quad 2\ Ag(NH_3)_2OH + 3\ H_2O$$

Reactions in the preparation of Tollens' reagent

$$\underset{\text{An aldehyde}}{R\overset{\overset{\displaystyle H}{|}}{-}C{=}O} + 2\ Ag(NH_3)_2OH \longrightarrow R\overset{\overset{\displaystyle O}{||}}{-}C{-}ONH_4 + 2\ Ag + NH_4OH$$

Tollens' reagent gives the best, and most spectacular, positive tests with formaldehyde and with sugars such as glucose and xylose. (Experiment 19) The reagent oxidizes higher molecular weight, water-insoluble aldehydes, but the silver precipitates as a black powder rather than as a silver mirror.

Caution!   *Silver nitrate and its solutions are toxic and leave dark stains on the skin. Exercise caution when working with these solutions.*

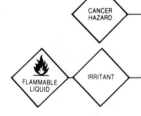

To prepare Tollens' reagent, clean four test tubes very carefully, using soap and water, and rinse them thoroughly with distilled water. To each test tube add 2 mL of a 5% solution of silver nitrate, one drop of 10% sodium hydroxide solution, and mix the ingredients thoroughly by swirling. Dilute 7 mL of concentrated ammonium hydroxide (28% ammonia) to 100 mL with distilled water. Add the dilute ammonia solution dropwise to each test tube with constant swirling or shaking, until the brown-gray precipitate of silver oxide *just dissolves*. The test will fail if you add a large excess of ammonia.

To one of four test tubes containing Tollens' reagent add a few drops of formaldehyde solution. (Note 4) Mix by shaking and warm the tube gently on the water bath. If the test tube is very clean and the reactants are not too concentrated, a silver mirror will form on the inner surface of the test tube. If the test tube is not clean, a black precipitate of finely divided silver will form. Repeat the test with acetaldehyde, benzaldehyde, and acetone. Record your results. (The silver mirror is easily removed from a glass surface with dilute nitric acid.)

Caution!   *Discard any unused Tollens' reagent after you have completed this portion of the experiment.* (Explosive silver nitrogen compounds form on prolonged standing.)

---

[1]Acetaldehyde boils at 21°C (70°F). In the laboratory it may be kept in an ice bath. Long term storage should be in an explosion-proof refrigerator.

2. **Fehling's solution**[1] is a solution of cupric hydroxide (stabilized as a tartrate complex). Benedict's solution[1] differs from Fehling's solution in that the cupric ion is stabilized as a citrate rather than as a tartrate. Both reagents are blue in color but are reduced by aldehydes to produce red cuprous oxide. In rare cases metallic copper is obtained. Neither reagent is affected by ketones. For simplicity, the copper complex may be represented as CuO (cupric oxide) in writing equations.

$$R-\overset{\overset{\displaystyle H}{|}}{C}=O + 2\,CuO + Na^+OH^- \longrightarrow R-\overset{\overset{\displaystyle O}{\|}}{C}-O^-Na^+ + Cu_2O + H_2O$$

An aldehyde   Sodium salt of
                                            an acid

Benedict's and Fehling's tests give the best positive tests with the lower molecular weight, water-soluble aldehydes and with sugars such as glucose and xylose. (Experiment 19) With the higher aliphatic aldehydes, careful observation may reveal some cuprous oxide formation at the interface between the reagent and the undissolved aldehyde. With some sugars, such as xylose, a copper mirror may form if the test is done carefully. These tests are negative for most aromatic aldehydes and for most ketones, but ketones with alpha hydroxy groups may react.

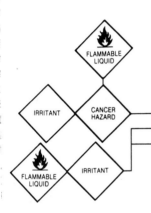

To each of four test tubes add either 5 mL of Benedict's solution or 5 mL of freshly prepared Fehling's solution (made by mixing 10 mL of Fehling's solution A and 10 mL of Fehling's solution B). To one tube add several drops of formaldehyde solution (Note 4), to the second add several drops of acetaldehyde, to the third add several drops of benzaldehyde, and to the fourth add several drops of acetone. Shake the test tubes *vigorously,* then place them in a boiling water bath 10–15 minutes. In a positive test the solution first turns a pale green, then a reddish precipitate of cuprous oxide forms. If no precipitate forms, add a few more drops of the carbonyl compound and continue heating as before. Record your observations.

3. **Schiff's fuchsin-aldehyde reagent**[2] is a sulfur dioxide-decolorized solution of the pink dye, fuchsin. Schiff's reagent reacts with aldehydes to produce a highly colored dye. The color produced usually is pink but may have a definite purple or blue cast. Ketones do not give a positive test when pure, but the Schiff's test is very sensitive and will be positive even if only a trace of aldehyde is present.

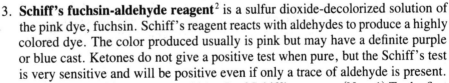

To each of five test tubes add 1 mL of Schiff's reagent. (Note 1) To the first test tube add a few drops of formaldehyde solution (Note 4), to the second a drop of acetaldehyde in 5 mL of water, to the third two drops of acetone in 5 mL of water, and to the fourth one drop of benzaldehyde in 5 ml of water. As a blank, add 1 mL of dilute sodium hydroxide to the fifth test tube. The latter treatment regenerates the original pink dye. Observe and record all color changes.

---

[1]See Instructor's Guide.
[2]Ibid.

4. The **Purpald® reagent** (Note 5) is a heterocyclic amino hydrazine that reacts with both aldehydes and ketones to give a cyclic intermediate. However, only the intermediate from an aldehyde can undergo air oxidation to give a highly colored product. The reaction proceeds as shown in the following equation:

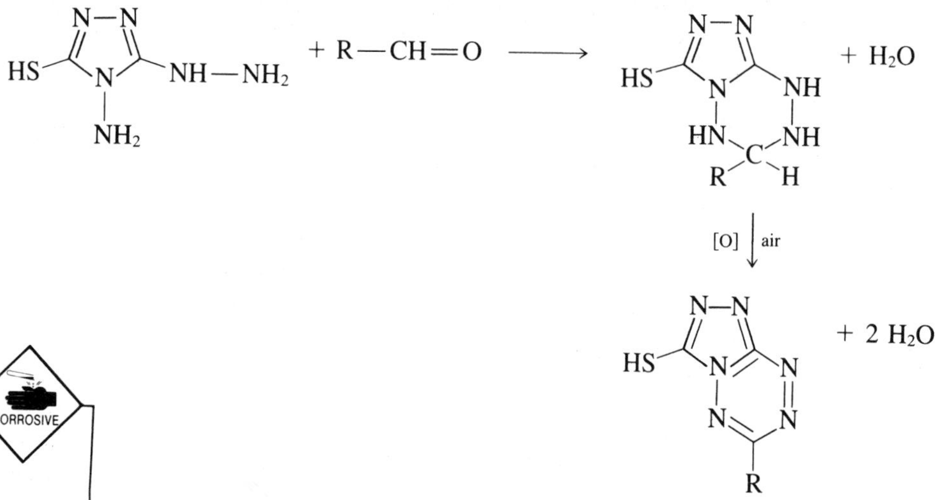

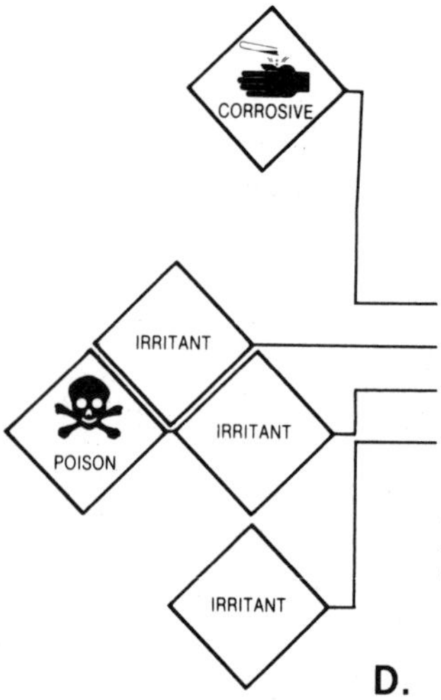

In each of four test tubes dissolve approximately 50 mg (a small mound about 5 mm in diameter) of Purpald® (Note 6) in 2 mL of 5% aqueous sodium hydroxide solution. Add 2 drops of heptaldehyde to the first test tube, 2 drops of benzaldehyde to the second, 2 drops of acetone to the third, and 2 drops of acetophenone to the fourth. Swirl and shake each tube vigorously so that the mixture is exposed to the air. In a positive test a deep purple color will develop, perhaps slowly at first but then becoming more and more intense. Record your observations.

D. Bisulfite Addition Product

Aldehydes and some ketones (methyl ketones and cyclic ketones) react with sodium bisulfite to yield bisulfite addition products.

$$R-\overset{\overset{\displaystyle H}{|}}{C}{=}O + NaHSO_3 \longrightarrow R-\overset{\overset{\displaystyle OH}{|}}{\underset{\underset{\displaystyle H}{|}}{C}}-SO_3^-Na^+$$

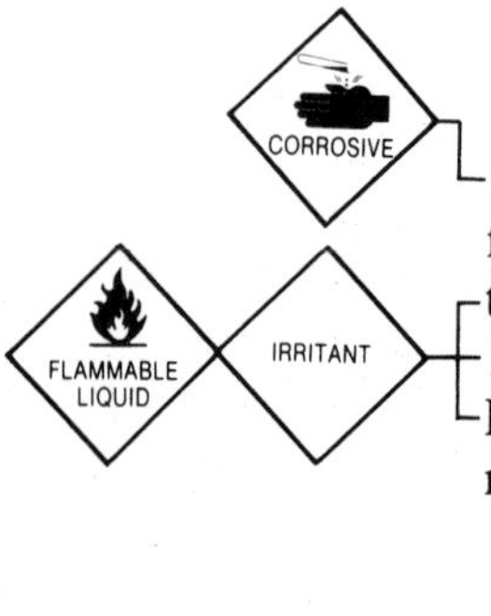

Place 5 mL of a *saturated* solution of sodium bisulfite in a 50-mL Erlenmeyer flask and cool the solution in an ice bath. Add 2.5 mL of acetone *dropwise*, swirling to mix the reactants. Allow the mixture to chill for about five minutes, then add 10 mL of ethyl alcohol. Stir the mixture and filter the crystals on the Hirsch funnel. Repeat the above procedure using cyclopentanone and diethyl ketone. Record your results.

Note 1 Prepare the reagent by dissolving 1.0 g of 2,4-dinitrophenylhydrazine in 5.0 mL of concentrated sulfuric acid. Slowly stir this solution into a mixture of 10 mL of water and 35 mL of 95% ethanol. Mix thoroughly.

Note 2 Prepare the reagent by dissolving 1.0 g of 2,4-dinitrophenylhydrazine in 10 mL of dimethylformamide.

Note 3 If the TLC part of the experiment is to be performed, the instructor should test the carbonyl compounds beforehand, as some ketones (for example, acetophenone) give excellent yields of 2,4-dinitrophenylhydrazones that do not perform well in this TLC system (because of poor solubility). Some good choices for TLC (and their approximate R*f* values) are: acetone (0.49), cyclopentanone (0.55), cyclohexanone (0.67), heptaldehyde (0.70), diethyl ketone (0.76), and methyl *tert*-butyl ketone (0.92).

Note 4 Prepare the formaldehyde solution by adding 1 mL of formalin solution to 20 mL of distilled water.

Note 5 Because of the sensitivity of the Schiff reagent to traces of aldehyde, the laboratory supply may be spoiled if students pour the reagent directly from the stock bottle into their aldehyde samples. Thus, students should obtain the necessary amount of reagent in a small flask and conduct all tests at their benches.

Note 6 Purpald® is the trademark of the Aldrich Chemical Company.

PART II   The Reduction of Carbonyl Compounds with Sodium Borohydride

Sodium (or potassium) borohydride is a highly selective reagent that reduces aldehydes or ketones to the corresponding alcohols but normally does not reduce nitro, nitrile, olefinic, amide, carboxylic acid, or ester functional groups. The reagent is less reactive (that is, more selective) than lithium aluminum hydride. Moreover, it may be used in aqueous or alcoholic solutions. Lithium aluminum hydride, on the other hand, reacts violently with such hydroxylic compounds and must be used in inert solvents such as ether and tetrahydrofuran. In this experiment a simple example of the use of sodium borohydride is illustrated by the reduction of benzophenone (diphenyl ketone) to benzhydrol (diphenylcarbinol).

Benzophenone

$$\left(\text{H} - \overset{\displaystyle\bigcirc}{\underset{\displaystyle\bigcirc}{\text{C}}} - \text{O} \right)_4 \text{BNa} + \text{HCl} + 3\,\text{H}_2\text{O} \longrightarrow$$

$$4 \; \overset{\text{OH}}{\underset{\text{H}}{\bigcirc - \text{C} - \bigcirc}} + \text{H}_3\text{BO}_3 + \text{NaCl}$$

Benzhydrol

PROCEDURE

Dissolve 4.5 g (0.025 mole) of benzophenone in 30 mL of methanol in a 150-mL beaker. In a smaller beaker prepare a solution of 1.0 g of sodium borohydride (**Caution!** *Flammable solid.*) in 15 mL of cold water. Stir the aqueous sodium borohydride solution in small portions into the benzophenone solution at such a rate that the temperature does not exceed 45°. The reaction is exothermic and the rate of addition, therefore, should not be too rapid. After all the sodium borohydride solution has been added, continue to stir the reaction mixture for approximately 15 minutes or until the diphenylcarbinol begins to precipitate. Continue to stir until a heavy slurry of crystals forms. Decompose the excess sodium borohydride by slowly stirring the crystalline slurry into a mixture of 100 g of crushed ice and water and 10 mL of concentrated hydrochloric acid in a 400-mL beaker. (Note 2) Collect the diphenylcarbinol on the Büchner funnel, wash the crystal cake twice with 50-mL portions of water, and dry. The product is of a high degree of purity but may be recrystallized from aqueous ethanol. Dry the product, weigh it, and determine its melting point. Yield, about 4.4 g (95%). Place your product in a properly labeled sample bottle and turn it in with your report.

Name   Section   Date

PART I Reactions of Aldehydes and Ketones

In the spaces below record after each substance tested the results obtained with each reagent used.

A. Test for Carbonyl Group

Carbonyl compound	Result

B. TLC of 2,4-Dinitrophenylhydrazones

Carbonyl compound	R_f

C. Distinguishing between Aldehydes and Ketones

Tollens' test

Carbonyl compound	Result

Benedict's test/Fehling's test

Carbonyl compound Result

_____ _____

_____ _____

_____ _____

_____ _____

Schiff's test

Carbonyl compound Color

_____ _____

_____ _____

_____ _____

_____ _____

Purpald® test

Carbonyl compound Color

_____ _____

_____ _____

_____ _____

_____ _____

D. Bisulfite Addition Products

Observations: _____

_____

_____

Name _____   Section _____   Date _____

PART II   Benzhydrol (Diphenylcarbinol)

Reaction Equation

$$(C_6H_5)_2CO \xrightarrow[\text{2. HCl, 3 H}_2\text{O}]{\text{1. NaBH}_4} (C_6H_5)_2CHOH$$

Benzophenone   Benzhydrol

Quantities   4.5 g   _____

Mol. Wt.   _____   _____

Moles   _____   _____

Equivalents   _____   _____

Theoretical yield _____ g

Actual yield _____ g

Percentage yield _____

mp _____ °C

1. Suggest a sequence of simple tests that will serve to distinguish between (a) benzaldehyde, (b) valeraldehyde, and (c) acetophenone.

2. Can you suggest a reason why diethyl ketone fails to form an addition product with sodium bisulfite while cyclopentanone gives an addition product?

3. Write equations for the reduction of the carbonyl groups in the following compounds with the appropriate metal hydride: (a) cyclohexanone, (b) n-valeric acid, (c) carbon dioxide, and (d) benzamide.

4. Write the equation for an alternative synthesis of benzhydrol from a carbonyl compound in which reduction is not involved.

*5. What singular structural feature of both aldehydes and ketones is most readily observed in the ir spectra of these classes of compounds?

*6. The formula C_4H_8O may be that of either a ketone or two aldehydes. How do the nmr spectra of these three compounds differ with respect to: (1) number of signals; (2) splitting of signals; (3) shift regions?

7. Plan a separation scheme (other than fractional distillation) by making a flowchart outlining the steps required to separate a mixture containing acetone, 2-propanol, and propionic acid.

Identification of an Unknown Carbonyl Compound

The formation of a crystalline derivative that is easy to purify and that possesses a sharp melting point is an excellent means of identifying an unknown organic compound. Such derivatives are especially helpful if the unknown is a liquid. The carbonyl compounds (aldehydes and ketones) form crystalline condensation products with a number of ammonia derivatives, but for identification purposes the hydrazones and the semicarbazones are most frequently prepared. Aldehydes and ketones react with phenylhydrazine to produce derivatives called **phenylhydrazones.** Derivatives formed from semicarbazide are called **semicarbazones.**

Phenylhydrazine   A phenylhydrazone

Semicarbazide   A semicarbazone

Often in qualitative work the prepared derivative of an unknown carbonyl compound is the **2,4-dinitrophenylhydrazone.**

2,4-Dinitrophenylhydrazine   A 2,4-dinitrophenylhydrazone

The substituted phenylhydrazone usually has a melting point higher than that of the unsubstituted phenylhydrazone. Moreover, certain carbonyl compounds form noncrystalline derivatives with phenylhydrazine. When this is the case, the 2,4-dinitrophenylhydrazone or the semicarbazone is prepared. The melting point of a phenylhydrazone, a 2,4-dinitrophenylhydrazone, or a semicarbazone, when compared with those recorded for carbonyl derivatives in tables and handbooks, often will serve to identify an unknown aldehyde or ketone. However, inasmuch as two or more different aldehydes and ketones may react with a reagent to give derivatives that melt at or near the same temperature, it frequently becomes necessary to prepare more than one derivative of a compound in order to identify it. It would be most unusual, however, for a compound to form two or more different derivatives with melting points near those of the corresponding derivatives of another substance. In this experiment the phenylhydrazones, the 2,4-dinitrophenylhydrazones, and the semi-carbazones of several carbonyl compounds are prepared. You then will be required to identify an unknown carbonyl compound.

> **Precautions.** *This experiment requires the use of the hydrazine derivatives phenylhydrazine, 2,4-dinitrophenylhydrazine, and semicarbazide. The strong mineral acids hydrochloric and sulfuric are also required. See HAZARD CATEGORIES 2 and 4, page 4.*

PROCEDURE   A.   Preparation of Phenylhydrazones

CANCER HAZARD

POISON

FLAMMABLE LIQUID

Dissolve 1 g of phenylhydrazine hydrochloride, $C_6H_5NHNH_2 \cdot HCl$, in 9 mL of warm distilled water. Add 1.5 g of crystalline sodium acetate, $CH_3COONa \cdot 3 H_2O$, and one drop of glacial acetic acid. If the solution is turbid, add a micro-spatula measure of charcoal, shake the solution vigorously, and filter. The solution deteriorates rapidly, so prepare it just before using. To the freshly prepared solution add 10 drops of benzaldehyde, C_6H_5CHO. Stopper the tube and shake vigorously until the product crystallizes. Filter the crystals with suction using the small Hirsch funnel and wash them thoroughly with water. Recrystallize the product from ethyl alcohol using the following procedure.

Transfer the crystals to a 50-mL Erlenmeyer flask, and add dropwise 1–2 mL of ethyl alcohol. Heat the mixture on the steam bath. If the product dissolves very readily in the initial amount of alcohol, add water to the alcoholic solution dropwise until a *faint* turbidity persists in the solution while it is hot. Add ethyl alcohol dropwise to the hot solution until it again becomes clear, then add one or two drops of alcohol in excess. Proceed as in the next paragraph. If the product does not dissolve completely in the initial volume of alcohol after boiling for a few minutes, add more ethyl alcohol dropwise while heating, repeating the process until all the solid dissolves. Use no more alcohol than necessary to effect solution of your product.

If the solution is dark or highly colored, add a small amount of charcoal and filter the solution. Cool the filtrate in the ice bath until the product crystallizes, filter the crystals with suction on the Hirsch funnel, wash them with a small volume (1 mL) of *cold* ethyl alcohol, and dry them as quickly as possible on a piece of filter paper.

Transfer a small amount of the product to a piece of clay plate or to a piece of filter paper on a watch glass, complete the drying by crushing the product on the plate (or paper), and determine the melting point.

B. Preparation of Semicarbazones

Dissolve 2 g of semicarbazide hydrochloride, H_2N—$NHCONH_2 \cdot HCl$, in 20 mL of distilled water. Add 3 g of crystalline sodium acetate. Mix and divide the reagent into two equal portions. Save one portion for Part D. To the other portion add 10 drops of methyl ethyl ketone. Stopper the test tube with a cork and shake vigorously. Filter the crystals using the Hirsch funnel and recrystallize from ethyl alcohol as described in Procedure A. When the product is dry, determine its melting point.

C. Preparation of 2,4-Dinitrophenylhydrazones

2,4-Dinitrophenylhydrazones may be prepared in sulfuric acid-ethanol solution, a time-honored method, or in N, N-dimethylformamide solution, a more recent (1984) method that is quite simple. Use the method for which your instructor provides the reagents.

(a) Sulfuric Acid Method

Dissolve 0.5 g of 2,4-dinitrophenylhydrazine in 3mL of concentrated sulfuric acid. Slowly stir this solution into 5 mL water and 15 mL of 95% ethyl alcohol. Mix thoroughly. Divide the mixture into *three* equal portions. Save one portion as before. To one portion stir in 10 drops of methyl ethyl ketone. The 2,4-dinitrophenyl-hydrazone usually forms immediately. If no precipitate results, set the reaction mixture aside for 15 minutes. Shake occasionally and scratch the wall of the test tube with a glass stirring rod to help induce crystallization. If the crystalline slurry that forms is too heavy to filter conveniently, add 10–15 mL of water. Prepare the 2,4-dinitrophenylhydrazone of benzaldehyde using the second portion of your re-agent and following the procedure just described. Separate and purify both deriva-tives according to the procedure described under Procedure A and determine their melting points.

(b) Dimethylformamide Method

Dissolve 0.6 g of 2,4-dinitrophenylhydrazine in 6 mL of N,N-dimethylformamide. Divide the solution into three equal portions. Save one portion as before. To one portion stir in 8 drops of methyl ethyl ketone; then add 2 drops of concentrated hydrochloric acid. Allow the reaction mixture to stand for 15 minutes, during which time some precipitate may form. Stir in 10 mL of 2*M* hydrochloric acid. Filter the precipitate on the Hirsch funnel with suction. Wash the precipitate in the following order: first with 10 mL of 2*M* hydrochloric acid to remove the unreacted 2,4-dinitrophenylhydrazine and the dimethylformamide, then with 20 mL of water to remove the acid, and finally with 2 mL of cold 95% ethanol. Recrystallize the product from ethyl acetate, using the minimum quantity of solvent. Effect solution by warming it on the steam bath. Cool the ethyl acetate solution in the ice bath. If the product does not begin to precipitate after standing for about 10 minutes, reheat it and add hexane dropwise while stirring. If no precipitate forms, add a volume of hexane equal to that of the ethyl acetate. Place the mixture in the ice bath. Shake occasionally and scratch the walls of the vessel with a glass stirring rod to induce crystallization. Filter and dry the product and determine its melting point. Prepare the 2,4-dinitrophenylhydrazone of benzaldehyde using the second portion of your reagent following the procedure just described.

Using this method some 2,4-dinitrophenylhydrazones (for example, that of cyclohexanone) precipitate immediately. These usually can be recrystallized from ethyl acetate without added hexane. Other 2,4-dinitrophenylhydrazones (for example, those of acetone and heptanal) are more soluble in dimethylformamide and do not precipitate until the dilute hydrochloric acid is added. These are usually quite soluble in ethyl acetate and require added hexane.

D. Identification of an Unknown Carbonyl Compound

Obtain 1 mL of an unknown carbonyl compound from your laboratory instructor. Prepare the 2,4-dinitrophenylhydrazone of your unknown, using the remaining portion of the reagent saved from Procedure C. Recrystallize the 2,4-dinitrophenylhydrazone and determine its melting point. Prepare the semicarbazone of your unknown, using the second portion of the reagent left from Procedure B. From Table 14.1 determine the identity of the unknown. If there is any doubt about the identity of the unknown *and* the phenylhydrazone of the unknown is a solid, you may prepare the phenylhydrazone as described in Procedure A and determine its melting point.

Table 14.1 Melting Points of Some Aldehyde and Ketone Derivatives (°C)

Compound and bp	Structure	Phenylhydrazone	Semicarbazone	2,4-Dinitrophenyl-hydrazone
Acetone (56)	$CH_3-\overset{O}{\overset{\|}{C}}-CH_3$	42	187	128
n-Butyraldehyde (74)	$CH_3CH_2CH_2CHO$	oil	104	123
Methyl ethyl ketone (80)	$CH_3-\overset{O}{\overset{\|}{C}}-C_2H_5$	oil	146	116
Diethyl ketone (102)	$C_2H_5-\overset{O}{\overset{\|}{C}}-C_2H_5$	oil	139	156
Methyl tert-butyl ketone (106)	$(CH_3)_3C-\overset{O}{\overset{\|}{C}}-CH_3$	oil	157	125
Cyclopentanone (131)	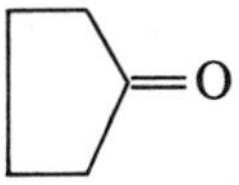 =O	50	205	142
Heptaldehyde (155)	$CH_3(CH_2)_5-\overset{O}{\overset{\|}{C}}-H$	oil	109	108
Cyclohexanone (155)	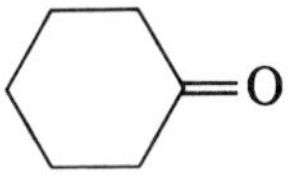 =O	77	166	162
Acetophenone (202)	$\overset{O}{\overset{\|}{C}}-CH_3$	105	199	239
Benzaldehyde (179)	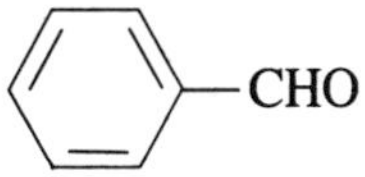 —CHO	158	222	237
Furfural (161)	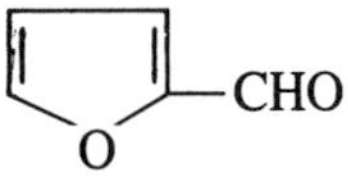 —CHO	97	202	229

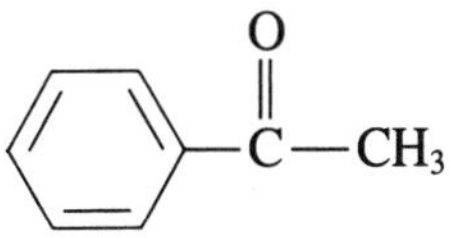

IRRITANT

FLAMMABLE LIQUID

Name _____   Section _____   Date _____

Derivatives of Aldehydes and Ketones

Compound	Phenylhydrazone	Semicarbazone	2,4-Dinitrophenyl-hydrazone
Benzaldehyde	mp _____ °C (experimental) mp _____ °C (Table 14.1)		mp _____ °C (experimental) mp _____ °C (Table 14.1)
Methyl ethyl ketone	(noncrystalline)	mp _____ °C (experimental) mp _____ °C (Table 14.1)	mp _____ °C (experimental) mp _____ °C (Table 14.1)

Identification of an Unknown Carbonyl

Number of unknown _____

Derivatives formed and melting points: _____

_____

Identity of unknown carbonyl compound: _____

1. A student has just enough of an unknown carbonyl compound to make one derivative and do one chemical test, provided that the test can be done with one drop of unknown. The student gets a good yield of a 2,4-dinitrophenylhydrazone, which melts at $125 \pm 3°C$. Based on a examination of Table 14.1, the student concludes that the unknown must be one of three possibilities. Fortunately, the student has done Part A(2), Part B, and all of Part C of Experiment 13 exactly as directed in this manual. With only one drop of unknown left, what additional test(s) can the student do that will identify the unknown?

2. Ethyl acetoacetate $(CH_3—\overset{\overset{\text{O}}{\|}}{C}—CH_2—\overset{\overset{\text{O}}{\|}}{C}—O—CH_2CH_3)$ reacts with phenylhydrazine under the usual conditions but gives a product with the molecular formula, $C_{10}H_{10}N_2O$. Suggest a structure for this product and write an equation for its formation.

Reactions of the Amines

The amines are the principal organic bases. Structurally, they are related to ammonia and have the general formulas RNH_2, R_2NH, and R_3N in which one, two, or all three of the hydrogen atoms of the ammonia molecule have been replaced by alkyl or aryl groups. Amines are classified as primary, secondary, or tertiary according to the number of hydrogen atoms of ammonia that have been replaced.

The amines are basic compounds because only three of the five electrons in the valence shell of the nitrogen atom are used in covalent bonding. An amino nitrogen with two unshared electrons can function as an electron-pair donor, or base. A water-soluble amine, like ammonia, may combine reversibly with water to produce a basic, substituted ammonium hydroxide according to the following equation.

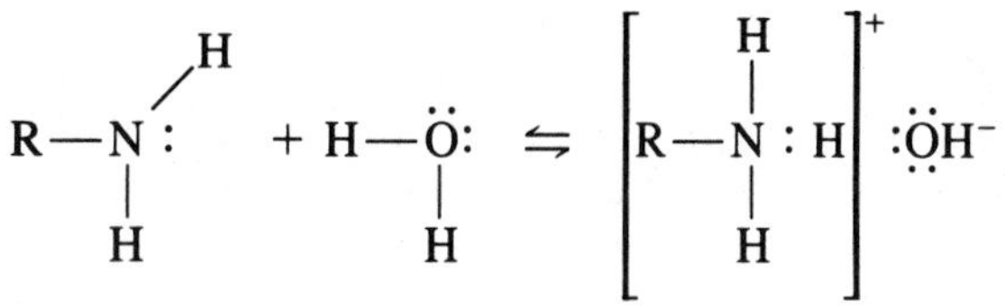

$$
R-\overset{\displaystyle H}{\underset{\displaystyle H}{N:}} \quad + \; H-\overset{\displaystyle}{\underset{\displaystyle H}{\ddot{O}:}} \; \rightleftharpoons \; \left[R-\overset{\displaystyle H}{\underset{\displaystyle H}{N:H}} \right]^{+} \; :\ddot{O}H^{-}
$$

Moreover, structures deficient by an electron pair (acids) may combine with an amine to produce salts.

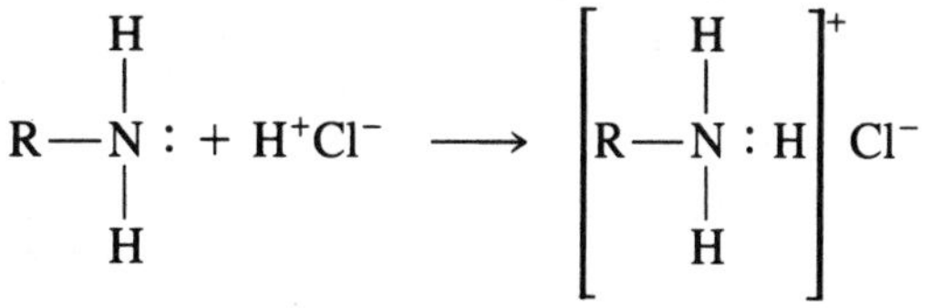

$$
R-\overset{\displaystyle H}{\underset{\displaystyle H}{N:}} + H^{+}Cl^{-} \longrightarrow \left[R-\overset{\displaystyle H}{\underset{\displaystyle H}{N:H}} \right]^{+} Cl^{-}
$$

The following series of experiments illustrate the properties of the amines.

Precautions. *This experiment requires the use of ether; primary, secondary, and tertiary amines; acid chlorides and anhydrides; phenyl isothiocyanate; the strong mineral acids hydrochloric and sulfuric; sodium hydroxide. See HAZARD CATEGORIES 1, 2, 3, 5 and 8, page 4.*

PROCEDURE   A.   Basic Properties of the Amines

Into four separate test tubes place small samples (about 10 drops each) of the following amines: aniline, benzylamine, diethylamine, and pyridine. To each test tube add 5 mL of water, stopper with a clean cork, and shake to mix. Are all amines completely soluble in water to give a homogeneous solution? Record your observations. Test each solution with red litmus paper. Are these solutions basic? Which of the solutions tested appears to be the strongest base?

Now add concentrated hydrochloric acid (use a dropper) to the test tubes containing the amine samples until each solution is distinctly acid to litmus. Have those amines that were not completely soluble in water now gone into solution? Write equations on your report form for the reactions that took place when HCl was added to each test tube. Mark for identification and save all solutions of amine hydrochlorides for Procedure C.

B.   Salt Formation

Place 10 mL of *anhydrous* ether in a clean, dry test tube and add 10 drops of aniline. Take the ether solution of aniline to the fume hood and bubble *dry* hydrogen chloride from the HCl generator (Fig. 15.1) through the solution for a few minutes until precipitation appears to be complete. Write the equation for this reaction on your report form. Collect the white precipitate by suction filtration on a small Hirsch funnel. Save a small amount for a melting point determination and transfer the balance to a clean, dry test tube. Add 5 mL of water. Result? Now test the aqueous solution with blue litmus paper. Also test the solution with a drop of silver nitrate solution. Now make the solution strongly basic with sodium hydroxide (use litmus). Shake and observe. Compare the appearance of the basic solution with that produced

---

Figure 15.1   Hydrogen chloride generator.

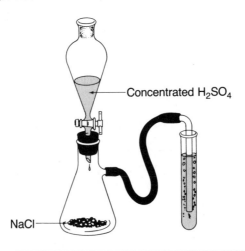

in the first part of Procedure A. Write equations for all reactions that have taken place.

C. Amines with Nitrous Acid

Nitrous acid is an unstable substance prepared in solution only when needed by reacting a mineral acid with sodium nitrite.

$$Na^+NO_2^- + H^+Cl^- \rightarrow Na^+Cl^- + HNO_2$$

When treated with nitrous acid a primary amine yields an alcohol as one product:

$$R-NH_2 + HNO_2 \longrightarrow R-OH + H_2O + N_2$$

The reaction is carried out by treating an acid solution of the amine (an amine salt solution) with an aqueous solution of sodium nitrite.

Secondary amines react with nitrous acid to produce neutral N-nitroso compounds. These are yellow oils that occasionally have some use in synthesis; however, they are known carcinogens and are currently the subject of study by cancer specialists. Small amounts of these dangerous materials exist in foodstuffs, beer, and Scotch whiskey, or are generated in our bodies. Nevertheless, we should avoid them in the laboratory.

$$\underset{R'}{\overset{R}{>}}N-H + H-O-N=O \longrightarrow \underset{R'}{\overset{R}{>}}N-N=O + H_2O$$

An N-nitroso
compound

Tertiary amines may simply dissolve in nitrous acid to form nitrite salts or undergo a rather complex degradation. The reaction is of little or no practical utility.

The reaction of aliphatic amines with nitrous acid has little value except as a diagnostic or as a separatory procedure. In either case, the primary amine is destroyed. On the other hand, the reaction of nitrous acid with aromatic primary amines produces intermediates known as diazonium salts.

$$\text{(NH}_2\text{ on benzene ring)} + 2\ HCl + NaNO_2 \xrightarrow{0-5°C} \text{(N}_2^+Cl^-\text{ on benzene ring)} + NaCl + H_2O$$

Benzenediazonium
chloride

A large number of very useful products that are difficult if not impossible to arrive at by any other route are synthesized by way of a diazonium salt. Many of our most useful dyes are azo dyes prepared in this way. (See Experiment 34, Procedure A.)

CANCER
HAZARD

Caution! *N-Nitrosoamines are carcinogenic. Do not treat the hydrochloride of diethylamine with sodium nitrite solution.*

Place the test tubes containing only the hydrochloride solutions of aniline, benzylamine, and pyridine prepared in Procedure A in a beaker of ice water. When the temperature has fallen to about 5° add to each test tube about 2 mL of a 5% solution of sodium nitrite. Observe closely and note if there is gas evolution from any of the test tubes. Remove the test tube containing the aniline sample and transfer it to a beaker of warm water (40°). Observe closely. Is there now any evolution of gas? Describe any change in appearance of the solution. Explain and write equations for the reactions that have taken place.

D.   Solid Derivatives of Amines

A convenient method for the identification of unknown amines is to convert them into sharp-melting crystalline derivatives. Most primary and secondary amines react with acetic anhydride to form substituted acetamides. This reaction is illustrated in the preparation of acetanilide (Experiment 30).

$$(CH_3CO)_2O + RNH_2 \longrightarrow CH_3CONHR + CH_3COOH$$

Acetic   N-substituted
anhydride   acetamide

Primary and secondary amines also form N-substituted benzamides but in this case the more reactive benzoyl chloride is used rather than the acid anhydride.

Benzoyl chloride   N-substituted benzamide

(a) Place 5 drops of benzylamine in a test tube and add 10 drops of water. Add to the aqueous solution of the amine 5 drops of benzoyl chloride and 2 mL of 10% sodium hydroxide. Stir with a stirring rod. The N-substituted benzamide should form almost immediately. Break up any lumps that form with your stirring rod, add 5 mL of water and filter. Wash with an additional 10 mL of water and recrystallize from 95% ethanol, dry, and determine the melting point of your derivative.

Most primary and secondary amines react readily with phenyl isocyanate and phenyl isothiocyanate to give substituted ureas and thioureas, respectively.

Phenyl isocyanate   N-substituted area

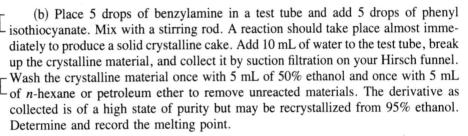

Phenyl isothiocyanate

The phenyl isothioureas are especially easy to prepare because, unlike phenyl isocyanate, phenyl isothiocyanate is not sensitive to water. The former combines with water to give an unstable carbamic acid that decomposes to aniline and carbon dioxide.

Caution! *Phenyl isothiocyanate is a lachrymator. Avoid contact with its vapors. Dispense and use this reagent in the exhaust hood.*

(b) Place 5 drops of benzylamine in a test tube and add 5 drops of phenyl isothiocyanate. Mix with a stirring rod. A reaction should take place almost immediately to produce a solid crystalline cake. Add 10 mL of water to the test tube, break up the crystalline material, and collect it by suction filtration on your Hirsch funnel. Wash the crystalline material once with 5 mL of 50% ethanol and once with 5 mL of *n*-hexane or petroleum ether to remove unreacted materials. The derivative as collected is of a high state of purity but may be recrystallized from 95% ethanol. Determine and record the melting point.

Name Section Date

REPORT 15

A. Basic Properties of Amines

Amine	Solubility in water	Action on pink litmus paper	Solubility in hydrochloric acid
Aniline			
Benzylamine			
Diethylamine			
Pyridine			

Strongest basic solution ______________________________

Reactions of amines with HCl:

B. Salt Formation

Aniline + dry HCl $\longrightarrow$

Melting point of product __________

Result upon adding water ______________________________

Equation:

Action of aqueous solution on blue litmus paper _____

Result upon adding a drop of $AgNO_3$ _____

Equation:

Result upon making solution basic _____

Equation:

Melting point of aniline hydrochloride _____

C.   Amines with Nitrous Acid

Complete the following reactions:

Aniline $+ HNO_2$ $\xrightarrow{0-5°}$

Aniline $+ HNO_2$ $\xrightarrow{40°}$

Explanation of color change _____

Equation:

Benzylamine $+ HNO_2$ $\xrightarrow{0-5°}$

Pyridine $+ HNO_2$ $\xrightarrow{0-5°}$

Name _____   *Section* _____   *Date* _____

D. Derivatives of the Amines

Equation for the reaction leading to a derivative of benzylamine with benzoyl chloride:

(a)

Melting point of substituted benzamide _____   _____
                                       Experimental   Literature

Equation for the reaction leading to a derivative of benzylamine with phenyl isothiocyanate:

(b)

Melting point of phenyl thiourea _____   _____
                                 Experimental   Literature

1. Outline a scheme that will lead to the isolation of a water insoluble amine when it is present in a mixture that also includes benzoic acid and anisole.

2. Considering the ring-activating influence of the amino group of aniline, suggest an easily prepared derivative of aniline in which only the hydrogen atoms of the ring are substituted.

3. The reaction of benzoyl chloride with primary and secondary amines in the presence of sodium hydroxide is referred to in textbooks as the **Schotten-Baumann reaction.** The concentration of hydroxide is a critical factor. A high concentration of sodium hydroxide dissolves the benzoyl derivatives of primary amines and is to be avoided. On the other hand, if the reaction medium were not slightly basic, a byproduct melting at 122° might coprecipitate with the substituted benzamide. What might this compound be?

4. Suggest a reason why the benzenesulfonamides,

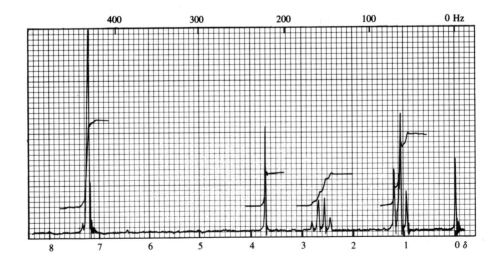

of primary amines are far more soluble in aqueous alkali than the corresponding benzamides.

*5. The N-H stretching frequency of the primary amino group usually appears as a doublet at 3200–3500 cm^{-1}. If an ir spectrum shows this absorption along with a strong absorption at or near 1700 cm^{-1}, what conclusions should you draw?

*6. An organic amine, $C_9H_{13}N$, gave the nmr spectrum below. Three substituted anilines and seven phenyl-substituted aliphatic amines correspond to this molecular formula. To which of the ten structures does the spectrum correspond?

7. Aniline and benzylamine both boil within the same temperature range (184–185°). How could one be distinguished from the other by a simple chemical test?

8. Consult a table of derivatives for primary amines and tell why the phenyl-thiourea derivatives would be of no help in identifying either compound in Question 7. What would be a good derivative in this case?

The Identification of an Unknown Organic Compound: Infrared and Nuclear Magnetic Resonance Spectroscopy

No other exercise in the organic laboratory program will test your knowledge of organic chemistry more completely or be more instructive than one requiring the identification of an unknown organic compound.

For many years the identification of unknown organic compounds as a laboratory exercise was deferred until students had reached a level of attainment and expertise that enabled them to carry out many chemical test reactions. However, the introduction of instrumental methods has greatly simplified the identification of an unknown organic substance. Many functional groups and other structural features of compounds now are readily identified by spectroscopic methods, and much of what at one time rested entirely upon the judgment of the student is accomplished by the infrared and the nuclear magnetic resonance spectrometers. These two instruments are of special value to the organic chemist in any identification scheme, but confirmation of the identity of an unknown should always be established by some supplementary chemical procedures.

By this time your laboratory program has included sufficient exercises dealing with the properties of the various classes of compounds to enable you to classify and identify any of the common organic substances listed in Table 16.1. In this experiment you will be issued one of the compounds listed in the table and you will be required to identify it. The ir and nmr spectra for each of the compounds in the table are identified by letter only and appear in the appendix. Your instructor will tell you which spectra are those of your unknown.[1] Study the spectra carefully and, with the help of Table 16.2 or one that appears in your text, try to derive as much structural information as you can that will enable you to place your unknown in its proper category—that is, among the acids, aldehydes, amines, alcohols, etc. From spectral studies alone it is possible to establish the identity of most of the compounds listed, but the practice of a little "wet" chemistry will help you make a positive identification. The next order of business, therefore, will be the accurate determination of a boiling point or a melting point of the compound. You should then test the solubility behavior of the unknown to help further classify it as an acid, an amine, or as one of the "neutral" compounds, such as aldehydes, ketones, esters, and amides. Finally,

---

[1] **Note to instructor.** A key to the identity of the compound corresponding to the spectra in the appendix is found on page 398.

you will want to carry out one or more of the simple chemical tests described in previous experiments that deal with the chemical properties of the different classes of compounds.

Record on the report form all the information from the above studies that has led to the identity of your unknown compound.

Precautions. *Until you have identified your unknown you will not know to which hazard categories it belongs. Therefore, treat all unknowns as if they were corrosive, carcinogenic, and toxic until you prove otherwise.* **Read Note 1, p. 198.**

Table 16.1   *Compounds to be Issued as Unknowns*

Compound	Structure	bp(°C)	mp(°C)
4-Chlorobenzaldehyde	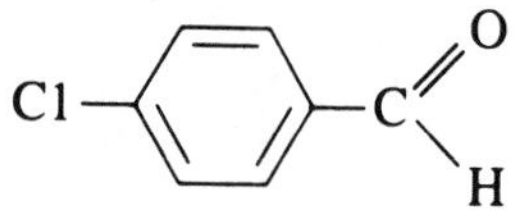		48
Benzhydrol	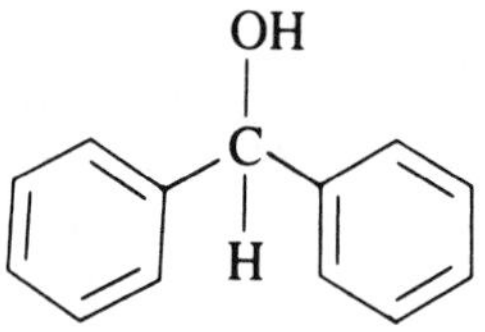		69
4-Chloroaniline	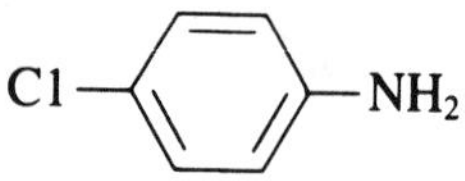		70
Ethyl acetate	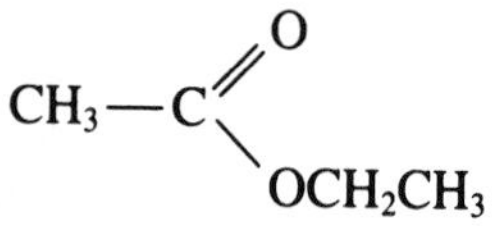	77	
2-Butanone	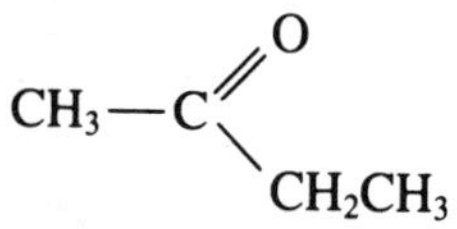	80	
2-Propanol	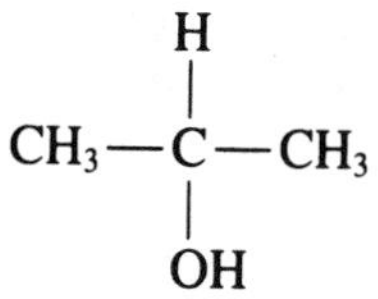	82	

Table 16.1 (Continued)   Compounds to be Issued as Unknowns

Compound	Structure	bp (°C)	mp (°C)
2-Methyl-2-propanol		83	
Propanol	$CH_3CH_2CH_2OH$	98	
3-Pentanone		102	
1,2-Dihydroxybenzene (Catechol)			106
Benzoic acid			122
o-Benzoylbenzoic acid			127
Benzamide			128
Cyclopentanone		131	
trans-Cinnamic acid			133

Table 16.1 (Continued)   Compounds to be Issued as Unknowns

Compound	Structure	bp(°C)	mp(°C)
Ethylbenzene	$-CH_2CH_3$	136	
Anisole	$-OCH_3$	153	
Benzaldehyde	$-C{\overset{O}{\underset{H}{}}}$	179	
Benzylamine	$-CH_2NH_2$	184	
N-Methylaniline	$-N{\overset{H}{\underset{CH_3}{}}}$	196	
Acetophenone	$-C{\overset{O}{\underset{CH_3}{}}}$	200	
m-Cresol	$HO-\quad-CH_3$	203	
Benzyl alcohol	$-CH_2OH$	204	
4-Ethoxyaniline (*p*-Phenetidine)	$NH_2-\quad-OCH_2CH_3$	250	

Table 16.2   A Simplified Correlation Table of IR Absorption Frequencies and NMR Proton Shift Values

Class of compound	Identifying group	Region of ir absorption (cm^{-1})	Chemical shift region (δ)	
			Ⓗ	Ⓗ
Acids		(a) 1680–1725 (b) 1250 (c) 2500–3000	10–12	2–2.6
Alcohols		(a) 1000–1200 (b) 3200–3600	1–7	3.4–4
Aldehydes		1690–1760	9–10	2–2.7
Amines		3200–3500 (—NH$_2$ may appear as a doublet)	1–5	
Aromatic rings		675–870 also 3000–3100	2.2–3	7–8
Ethers		(a) 1200–1275 (b) 1060–1150		3.3–4
Ketones		approx. same as aldehydes		approx. same as aldehydes
Phenols		(a) 1140–1230 (b) 3200–3600	4–12	

Note 1 Although boiling points are provided in Table 16.1 in the event that you wish to use them for identification purposes, do not attempt to take a boiling point of your unknown unless your instructor has told you that it is safe to do so. If you are permitted to determine the boiling point, the procedure described on pages 148–150 is recommended. (**Instructor: Read Note 1, p. 57.**)

Name _____ Section _____ Date _____

Unknown No.: _____

1. Conclusions drawn from ir and nmr spectra studies:

Significant ir absorption frequencies (cm^{-1})	Inferences
Significant nmr chemical shift values (δ)	Inferences

2. Physical constants: mp _____ °C

   bp _____ °C

3. Solubility tests:

H_2O	Ether	10% HCl	10% NaOH	10% Na_2CO_3	85% H_3PO_4

4. Chemical classification tests:

Reagent	Results	Inferences

5. Special tests: (describe)

Identity of unknown compound: _____

*1. The carbonyl group shows an ir absorption somewhere within the wavelength range of 5.347–6.134 μ. Show how these wavelength values may be expressed in terms of wave numbers within the range of 1870–1630 cm^{-1} respectively.

*2. In the explanation of ir absorption phenomena, bonded atoms are frequently pictured as small spheres connected by miniature springs, and vibrational frequencies of the bonded atoms likened to that of a simple harmonic oscillator. Refer to your text and, from the Hooke's law equation, tell what mainly determines the characteristic absorption frequency of any functional group.

*3. An organic compound, $C_9H_{10}O$, may be either an aldehyde or a ketone. What single feature in the nmr spectra of one of these two classes of compounds easily distinguishes it from the other?

*4. If, on strong oxidation with chromic acid, the compound in Question 3 yielded only benzoic acid, how many isomeric structures could be drawn with this molecular formula? If the nmr spectrum of the compound showed only three singlets, what is its structure and name?

5. If you were issued an unknown carbonyl compound, $C_6H_{12}O$, and asked to identify it, outline in a stepwise procedure exactly how you would go about it. If necessary, use more than the four spaces provided.

Operation	Result	Inference
(1)		
(2)		
(3)		
(4)		

6. Using the *Aldrich Catalog Handbook of Fine Chemicals*, prepare a table like Table 16.1 which lists the name, structure, and hazard categories for all of the compounds in Table 16.1.

Fats and Oils;
Soaps and Detergents

The fats and oils of animal and vegetable origin are **glycerides** or long-chain fatty acid esters of the trihydric alcohol glycerol, $HOCH_2CH(OH)CH_2OH$.

$$CH_2O-COR$$
$$|$$
$$CH-O-COR'$$
$$|$$
$$CH_2O-COR''$$

A general formula for a fat or oil

The hydrocarbon segments indicated by R, R′, and R″ in the formula above generally are different. They may vary not only in length but also in degree of unsaturation. If the fatty acid components of a glyceride are long-chain ($C_{12}-C_{18}$) and saturated, the ester is a solid or a semisolid at room temperature and is classified as a fat. If the long-chain acid residues are unsaturated (that is, contain one or more double bonds), the glyceride is a liquid at room temperature and is classified as an oil. Unsaturation in the acid components of a fat lowers its melting point. Conversely, saturating the double bonds with hydrogen raises its melting point. The latter process, when applied to oils, is known as hardening or hydrogenation and is carried out as an industrial process for the manufacture of margarines and cooking fats from vegetable oils.

A fat or an oil, when saponified (that is, hydrolyzed with alkali), produces glycerol and the sodium or potassium salts of a mixture of fatty acids. The latter are called **soaps.**

$$
\begin{array}{ll}
CH_2O-COR & \qquad CH_2OH \quad RCOO^- \, Na^+ \\
| & \qquad\quad | \\
CH-O-COR' + 3\,NaOH \longrightarrow & CHOH + R'COO^- \, Na^+ \text{ (soaps)} \\
| & \qquad\quad | \\
CH_2O-COR'' & \qquad CH_2OH \quad R''COO^- \, Na^+ \\
\text{A fat or oil} & \qquad\qquad\quad \text{Glycerol}
\end{array}
$$

The sodium and potassium soaps are soluble in water and are used as cleansing agents. The calcium, magnesium and ferric salts of the same fatty acids are insoluble in water and are not useful as soaps. These insoluble metal salts precipitate as a scum when ordinary soap is used in hard water.

Synthetic detergents, also called **syndets** or **surfactants,** resemble soaps in that most of them have a long, nonpolar, hydrophobic (water-hating) hydrocarbon tail and a polar, hydrophilic (water-loving) head. However, the polar head may be anionic, cationic, or neutral. Some typical synthetic detergents are shown below.

$$CH_3(CH_2)_m - CH - (CH_2)_n CH_3$$

$$CH_3(CH_2)_{11}OSO_3^- Na^+ \qquad R - N(CH_3)_3^+ \ Cl^-$$

$$SO_2O^- Na^+$$

An alkyl benzene sulfonate   Sodium lauryl sulfate   A quaternary ammonium salt
(Anionic: m+n = 9–12)   (Anionic: a common syndet)   (Cationic: R = C_{12}–C_{16})

$$R - \langle\!\!\!\bigcirc\!\!\!\rangle - O - (CH_2CH_2O)_n CH_2CH_2OH \qquad R - \overset{O}{\overset{\|}{C}} - OCH_2 - \overset{CH_2OH}{\underset{CH_2OH}{\overset{|}{\underset{|}{C}}}} - CH_2OH$$

A nonionic detergent   A nonionic detergent
(R = C_8–C_{10}; n = 8–12)   (R = C_{11}–C_{17})

Usually synthetic detergents do not form insoluble salts with the metallic ions normally present in water.

Although soaps are water-soluble, the long-chain fatty acid from which they are formed are not. Therefore, acidification of a soap solution will cause the fatty acid to precipitate.

$$RCOO^- Na^+ + HCl \longrightarrow RCOOH + Na^+ Cl^-$$
$$\downarrow$$

Soap   Fatty acid

Cationic detergents are generally soluble in acidic solution; however, the behavior of other types is less predictable.

 Precautions.   *This experiment requires the use of ethanol, methylene chloride, bromine, sodium hydroxide and the strong mineral acids hydrochloric and sulfuric. See HAZARD CATEGORIES 1, 2, and 3, page 4.*

PROCEDURE   **A.   Solubility**

Note the odor and appearance of cottonseed oil. Try dissolving a little in water, in alcohol, and in methylene chloride. Repeat the test using samples of margarine.

B. Unsaturation Tests

Dissolve 0.5 mL of cottonseed oil in 5 mL of methylene chloride in a test tube. Add a solution of 5% bromine in methylene chloride dropwise, counting the number of drops required until the bromine color no longer is discharged instantly. Repeat the test using 0.5 g of Crisco or some other hydrogenated shortening.

C. Drying Oils

On different parts of a glass cover plate place one drop of each of the following oils: (a) boiled linseed oil, (b) cottonseed oil, (c) Mazola or some other corn oil. Put an identifying mark by each and let them stand in your locker until the next laboratory period. Observe and report the condition of each oil at that time.

D. The Preparation of Soap

In this part of the experiment a boiling water bath is used as a heat source. Heat the bath either by a burner or by an electric hot plate. If you are using a burner, support the water bath filled about two-thirds with water on an iron ring or tripod. Heat the water to boiling. Dissolve 2.5 g of sodium hydroxide in 5 mL of distilled water and 10 mL of ordinary alcohol (95%). Add the alkaline solution to 5 g of Crisco in a 150-mL beaker. Cover the beaker with a watch glass and heat the mixture on the water bath. Stir frequently to prevent spattering and keep the volume of the solution fairly constant by adding small amounts of 50% alcohol. If the mixture foams too much add a small amount of undiluted alcohol. The reaction is complete when the oil or melted fat has dissolved and gives a clear homogeneous solution (about 30 minutes). Dilute your soap solution by adding 15 mL of water; then pour it into a brine made by dissolving 30 g of sodium chloride in 100 mL of distilled water. Stir the mixture thoroughly and collect the precipitated soap on the Büchner funnel. Wash the soap twice with 10-mL portions of cold distilled water.

Dissolve 2 g of the crude soap in 100 mL of distilled water and set the mixture aside as your test solution. Place the remainder of the soap in an evaporating dish, heat it on the water bath, and stir into the soap just enough water to form a thick solution. Allow the soap solution to cool. Unless the amount of water added was excessive, the soap will solidify into a cake somewhat resembling commercial soap.

Perform the following experiments on samples of your test solution and record your results on the report form.

(a) Alkalinity

Test the alkalinity of your dilute soap solution with pink litmus paper and compare the result with those of similar tests made on solutions prepared from 1.0 g samples of household soaps or detergents, each dissolved in 100 mL of distilled water. Record the names of the detergents used and your results.

(b) Metallic Salts of Fatty Acids

To a 10-mL portion of your soap solution add 1 mL of a dilute (0.1%) calcium chloride solution. Shake vigorously and observe. Repeat the test with dilute magnesium chloride and ferric chloride solutions. Perform the same tests on the samples of household detergent solutions prepared in (a). Record your results.

(c) Precipitation of Soap

To 50 mL of your soap solution add dilute hydrochloric acid dropwise until the solution is acid to Congo Red paper. Cool the mixture in ice, collect the precipitated acid on your suction funnel, and wash it with 20 mL of cold water. On your report form write an equation to show the reaction that took place. Assign a name to your product and show it to your laboratory instructor when handing in your written report.

Test the solubility of a small sample (0.25 g) of your product in 2 mL of methylene chloride. Also test the solubility of a small sample of stearic acid in the same solvent.

(d) Emulsifying Action of Soap

Shake 4 drops of mineral oil with 10 mL of soap solution. Repeat the experiment using 10 mL of water and 10-mL samples of the household detergent solutions. Reexamine all mixtures after they have been standing for 5 minutes. Record your results.

REPORT **17**

Name ____________ Section ____________ Date ____________

Fats and Oils; Soaps and Detergents

Solubility of Fats and Oils

Solvent used	Observation
Water	
Alcohol	
CH_2Cl_2	

Unsaturation Tests

Number of drops 5% bromine in methylene chloride required to saturate your sample of cottonseed oil. ________

Number of drops of 5% bromine in methylene chloride required to saturate a sample of Crisco. ________

Draw the structure of the principal glyceride found in cottonseed oil (consult text) and write the equation for the reaction that took place with bromine.

Drying Oils

Describe the appearance of your oil samples after standing one week.

Boiled linseed oil. ____________________

Cottonseed oil. ____________________

Mazola or other corn oil. ____________________

Soaps and Detergents

Tests performed	Your soap (brand X)	Detergents	
(a) Alkalinity (litmus)			
(b) CaCl₂ solution			
MgCl₂ solution			
FeCl₃ solution			

(a) Equation for reaction with HCl

Solubility of product in CH_2Cl_2. _____

Solubility of stearic acid in CH_2Cl_2. _____

(b) Emulsifying action of soap on oil.

Combination	Results
Oil and water mixture	
Oil and soap mixture	
Oil and _____ mixture	
Oil and _____ mixture	

QUESTIONS
and
EXERCISES

1. Name a glyceride that probably is present in lard; in Mazola oil.

2. Could cottonseed oil be used in paints as a drying oil? Explain.

3. Assume the average molecular weight of a fat is 890. What fatty acid components probably predominate?

4. Why was the dilute soap solution poured in a solution of sodium chloride in water?

5. How does soap function as an emulsifying agent for oil in water? Draw a picture of an oil droplet and several soap molecules to illustrate your answer.

6. Suppose that you wanted to emulsify a water-insoluble compound in water. Would soap be a reasonable choice for the emulsifying agent if the water were slightly acidic? Why? If it would not be a good choice, can you draw the structure of an organic molecule that might be more suitable?

7. If castor oil (Experiment 33) were treated with sulfuric acid and then neutralized with sodium hydroxide, what would probably be the result? Would this structure be water-soluble?

8. Saponification (conversion into soap) is a term also used to describe the alkaline hydrolysis of an ester. Why is alkaline hydrolysis of an ester preferred over acid hydrolysis?

9. Examine the labels on the synthetic detergents used in this experiment and make a list of the detergents or surfactants in each. Do the same for any soaps or detergents that you can find in your home. (Note: Some manufacturers are more informative than others as to the ingredients in their products.)

Stereochemistry

Two or more organic compounds having the same molecular formula but different structures are said to be **isomers.** Those isomers that have the same carbon skeleton and the same substituents in the same locations (that is, have the same sequence of covalent bonds) but differ in the relative disposition of their atoms in space are said to be **stereoisomers.** For convenience we can divide those substances existing in stereoisomeric forms into two broad classes: *cis-trans* **isomers** and **optical isomers.** The *cis-trans* isomers include those isomers whose only differences lie in the arrangement of the four substituents on a carbon-carbon double bond or in the relative arrangement of at least two substituents on tetrahedral atoms in a cyclic structure. There are several subtypes of optical isomers, but the type in which we are most interested consists of those isomers that differ from one another in the disposition of substituents about one or more chiral centers, usually "chiral carbon atoms" (or "asymmetric carbon atoms"). Often such optical isomers exist as pairs of **enantiomers,** in which one enantiomer is the mirror image of the other. Enantiomers rotate the plane of plane-polarized light in opposite directions and can often be characterized on the basis of their **optical rotation;** that is, they are said to be optically active. Optical isomers that are not mirror-image related also exist and are called **diastereomers.** Diastereomers may or may not show optical activity. Many of the most important biological molecules exist largely, if not entirely, in one enantiomeric form. Because of this, at any one time there exists on earth a vast collection of more or less "pure" enantiomers, that is, not mixtures of mirror-image twins. This collection is called the **chiral pool** and is a resource from which we can draw when we need a starting material to synthesize an enantiomerically pure substance and to which we contribute when we synthesize a single enantiomer or separate two enantiomers. *Cis-trans* and optical isomerism are discussed in some detail in your text, and you should review that discussion in order to place this experiment in the proper perspective.

The first part of this experiment illustrates the isomerization of a *cis*-alkene into a *trans*-alkene and shows some of the often pronounced differences in physical properties of such isomers. The second part of this experiment illustrates the use of an enantiomer taken from the chiral pool to synthesize an enantiomer of a different substance. Of course, in this experiment we borrow one substance from the chiral pool and return another. The third part of this experiment illustrates the use of a biological catalyst, an enzyme system, to convert a nonchiral molecule into one enantiomer of a chiral molecule, thereby adding to the chiral pool.

PART I The Isomerization of Maleic Acid to Fumaric Acid

Butene-1,2-dioic acid exists in both *cis*-, and *trans*-forms. For historical reasons the two isomers have been given different names: the *cis*- form is named maleic acid and the *trans*-form is named fumaric acid.

Maleic acid

Fumaric acid

Generally, the more stable of two simply substituted alkenes is the *trans*-isomer. In this experiment we take advantage of this difference in stability to isomerize the less stable *cis*-isomer, maleic acid, into the more stable *trans*-isomer, fumaric acid. The isomerization has been achieved using several techniques. Here you will use the simplest of these techniques: reaction with hot hydrochloric acid. Note that this procedure is specific to this isomerization and is not a general method for the isomerization of *cis*- to *trans*-isomers.

Precautions.   *This experiment requires the use of ether, hexane, the organic acids maleic and fumaric, and the mineral acids hydrochloric and sulfuric. See HAZARD CATEGORIES 1, 2, and 8, page 4.*

PROCEDURE

Arrange a reflux apparatus such as one of those illustrated in Figures 0.1, 0.2, or 0.3, using a 50-mL round-bottomed flask. Because the solvent in this reaction is water, any of the sources of heat should be acceptable. To the round-bottomed flask add 4.0 g of maleic acid, 10 mL of water, and 10 mL of concentrated hydrochloric acid. Swirl the flask to mix the reactants, add one or two Boileezers, and reassemble the apparatus. Heat the reaction mixture under reflux for 20 minutes, during which time fumaric acid should begin to precipitate. Remove the source of heat and, as soon as the flask is cool enough to handle, remove it and cool the reaction mixture in the ice bath for 5 minutes. Filter the product with suction on a small Büchner funnel and wash it with 20 mL of ice-cold water. Transfer the product to a 250-mL Erlenmeyer flask, and recrystallize it from 70 mL of water by heating it on the hot plate or over a burner on a ceramic-fiber centered wire gauze. Cool the aqueous solution in ice; then collect the purified fumaric acid on the Büchner funnel. As fumaric acid melts at 299–300°, do not attempt to take its melting point unless your instructor tells you to do so. (Note 1)

Note 1   Take melting points in a metal block device with sealed capillaries.

PART II   Preparation of L-3-Phenyllactic Acid

Although simple aliphatic primary amines react with nitrous acid to give mixtures of products resulting from the decomposition of intermediate alkane-diazonium ion, α-amino acids react much more cleanly to form an α-hydroxy acid and nitrogen. This reaction has been used most frequently as a test for the presence of free amino groups in polypeptides and proteins called the van Slyke method. However, it can be used to prepare α-hydroxy acids from α-amino acids and has the interesting feature that the stereochemistry of the product is the same as that of the starting material. Thus, in this reaction we may convert one member of the chiral pool into another of different gross structure but of the same relative configuration. In this experiment you will convert L-phenylalanine into L-3-phenyllactic acid, which has been utilized in biochemistry as an enzyme inhibitor.

$$C_6H_5-CH_2 \quad COOH \xrightarrow{\text{HONO}} C_6H_5-CH_2 \quad COOH$$

L-Phenylalanine $\longrightarrow$ L-3-Phenyllactic acid

PROCEDURE   ***Caution!***   *Since oxides of nitrogen can evolve in this experiment if it is done carelessly, it may be prudent to carry out the experiment in the exhaust hood.*

Dissolve 3.30 g of L-phenylalanine in 125 mL of hot distilled water in a 250-mL Erlenmeyer flask, either on the hot plate or over a ceramic-fiber centered wire gauze heated by a low flame. The amino acid is not very soluble in cold water but will dissolve easily in the specified amount of water. As soon as the amino acid dissolves, remove the heat source and allow the flask to cool to room temperature. Meanwhile, in one 50-mL Erlenmeyer flask prepare a solution of 1.66 g of sodium nitrite in 25 mL of water and in another 50-mL Erlenmeyer flask prepare or obtain 25 mL of 1N sulfuric acid. If the temperature of the amino acid solution is still above 30°, put it carefully in cool water. Do not overcool the solution or the amino acid may precipitate. (Note 1) If a magnetic stirrer is available, place the stirrer bar in the cooled amino acid solution and start the stirrer at a low speed. (Note 2) Over a period of about 30 minutes add the sodium nitrite and sulfuric acid solutions, using two Pasteur pipets and a drop rate of about one drop per second. (Note 3) Each time you refill the pipets, swirl the reaction mixture if you are not using a magnetic stirrer. If more than a trace of the brown fumes of oxides of nitrogen appears above the solution, you are probably adding the solutions too rapidly or stirring inadequately. After the addition of the two solutions is complete, loosely cork the reaction vessel and allow the mixture to stand (preferably in the hood) until the next laboratory period.

Transfer the reaction mixture to a 250-mL separatory funnel and extract it three times with 25-mL portions of ether. (Note 4) Combine the ether extracts and dry them over about 1 g of magnesium sulfate for about 15 minutes. Filter the solution through a small plug of cotton in a conical funnel into a 100-mL (or larger if necessary) round-bottomed flask. Add a Boileezer and incorporate the flask into a simple distillation apparatus (Fig. 3.1) and distill off the ether using either a hot-water bath (**Caution!** *No flames!*) (Note 4) or a heating mantle. Collect the distillate in an ice-cooled receiver. Be ready to lower the heat source if the ether begins to distill too rapidly. When the volume of solution in the still pot has been reduced to about 15–20 mL, stop the distillation. Transfer the solution in the still pot to a 125-mL Erlenmeyer flask. Rinse the still pot with 5–10 mL of the ether you removed by distillation and add the rinsings to the Erlenmeyer flask. Pour the remainder of the ether distillate into the waste ether receptacle provided for this purpose. Dilute the solution in the Erlenmeyer flask with 50 mL of hexane (up to 75 mL if you have more than about 20 mL of solution). Chill the ether-hexane mixture in the ice bath for 15–20 minutes, occasionally scratching the walls with a glass stirring rod if necessary to induce crystals to form. Filter the mixture with a small Büchner funnel and wash the crystals with a little hexane. Allow the product to air-dry, then weigh it and determine its melting point. Expected yield, 1.5–2.5 g; mp 122–125°. If a suitable polarimeter is available, determine the optical rotation of your product. (Note 5)

Note 1 If only a small amount of the amino acid precipitates, it is not necessary to try to bring it back into solution. If considerable precipitate forms, add another 20 mL of water and reheat.

Note 2 The yield in this reaction apparently depends on the concentration of reactants, the pH, the temperature, and perhaps other factors as well. The best yields are obtained when the sodium nitrite and sulfuric acid solutions are added slowly to the stirred solution of phenylalanine at the lowest optimum temperature. However, even if all of these conditions are not properly controlled, you should obtain a yield in the range of 40–50%. The solution can be added by dropping funnel or by buret, but control of the drop rate with these is no easier than with Pasteur pipets.

Note 3 If a magnetic stirrer is available, you may use a round-bottomed flask instead of the Erlenmeyer flask to facilitate the action of the stirrer.

Note 4 The quantity of ether involved can be removed with one filling of hot water in an average-sized metal water bath. You can heat the water on a hot plate in one hood set aside for this purpose.

Note 5 Measured specific rotations for this product fall in the range $[\alpha]_D = -20$ to $-21°$ ($C = 2.0$ water). These values correspond to an observed rotation of only about $-0.4°$ for a 1-dm polarimeter tube; therefore, the polarimeter must be able to handle this small value.

PART III The Enzymatic Reduction of Ethyl Acetoacetate by Yeast: (S)-(+)-Ethyl 3-Hydroxybutanoate

The laboratory preparation of an organic compound that contains a chiral center normally results in a mixture of equal amounts of enantiomeric forms because the atom that is to become the chiral center may be approached by the attacking reagent at random from one direction or another. Inasmuch as countless molecules are involved, the probability of producing a (+)-isomer is the same as that of producing a (−)-isomer. Such 50-50 mixtures of equal parts of enantiomorphs are called **racemic mixtures** and are optically inactive because the optical activity of one half of the mixture is nullified by that of the other half.

Nature, on the other hand, can be very selective in biochemical changes because the attacking reagent itself is chiral in most cases and a molecule of the substrate may be selectively attacked in a stereospecific manner. The formation of one enantiomer thus is favored to the exclusion of the other if the reaction is one that results in a chiral product. For example, the catalytic reduction of pyruvic acid, $CH_3COCOOH$, results in racemic lactic acid, $d, l\text{-}CH_3CH(OH)COOH$, but reduction by *reductase*, which is present in yeast, yields only (−)-lactic acid.

The following experiment illustrates the selective reduction of ethyl acetoacetate to (S)-(+)-ethyl 3-hydroxybutanoate by yeast in a sucrose solution.

$$CH_3-\overset{\overset{\textstyle O}{\|}}{C}-\overset{\overset{\textstyle H}{|}}{\underset{\underset{\textstyle H}{|}}{C}}-\overset{\overset{\textstyle O}{\|}}{C}-OC_2H_5 \xrightarrow[\substack{H_2O,\ \text{sucrose} \\ 25-30°}]{\text{yeast}}$$

Ethyl acetoacetate

(S)-(+)-Ethyl 3-hydroxybutanoate

PROCEDURE

IRRITANT

In a 2-liter beaker equipped with a stirrer dissolve 100 g of sucrose in 400 mL of warm (35°) water. To the warm sugar solution add 35 g of dry baker's yeast. Stir to thoroughly mix the yeast and sugar solution. Fermentation will begin within a few minutes. To the warm fermenting mixture add 12.5 mL (13 g, 0.1 mole) of ethyl acetoacetate in small portions. After the ester has been added, keep the mixture uncovered and allow it to stand in a warm (25–30°) place for a period of 72 hours or longer. During the standing time the mixture should be stirred or agitated frequently. (Note 1)

Prepare a filter base by pouring a slurry of Celite into a 10-cm Büchner funnel. Discard the filtered water, reassemble the funnel and flask, and filter the fermented mixture by carefully decanting the supernatant liquid into the funnel. After filtering the solution, add 50 mL of warm water to the residual yeast, agitate, and filter. Add this filtrate to the first. (Note 2) Saturate the filtrate with sodium chloride and extract it four times using 50-mL volumes of ether for each extraction. (Note 3) Add 10 g of anhydrous magnesium sulfate to the combined extracts or enough to flow along the bottom of the flask without forming lumps. While the ether solution is drying, set up a distillation apparatus such as that illustrated in Figure 3.1. Transfer the dried ether extracts to the distillation flask via a long-stemmed funnel fitted with a small cotton plug, add a Boileezer, and remove the ether by distillation. Use a heating mantle or a warm-water bath (***Caution!*** *No flames.*). When the volume is reduced to approximately 10 mL and no more ether appears to collect in the condenser, stop heating. Return the collected ether to the bottle labeled "Solvent ether from extractions." Transfer the residue to a smaller distillation flask, attach a clean, small receiver, and distill the residue under diminished pressure. Use a heating mantle as your heat source. Collect that portion boiling at 71–73° (12 mm). (Note 4) Make the following determinations:

1. Test a drop of your product with a 1.0% ferric chloride solution to determine if unchanged ethyl acetoacetate is present (see Experiment 12, Part II-C).

2. Determine the specific rotation, $[\alpha]_D^{25}$, of your product by measuring the optical rotation using methylene chloride as solvent (see Experiment 9). The specific rotation of enantiomerically pure (*S*)-(+)-ethyl 3-hydroxybutanoate is reported to be +43.5 in chloroform.

3. Report your **optical yield.** The optical yield is a measure of optical purity. For example, a specific rotation $[\alpha]_D^{25}$ of +39.15 corresponds to an optical yield or **enantiomeric excess** of 90%.

4. Obtain an ir spectrum of your product and compare it with that given by ethyl acetoacetate.

5. Obtain an nmr spectrum and compare it with that for ethyl acetoacetate.

Note 1   Stirring the fermenting mixture, keeping it warm, and permitting it to stand for a long period of time are critical factors in the success of this enzymatic reduction.

Note 2   The filtration step will be very slow if the yeast is allowed into the funnel at the start of filtration. The authors have found it expedient to siphon the supernatant liquid from the spent yeast first, then filter the yeast suspension.

Note 3   There may be some tendency for emulsion formation if the ether and aqueous solution are shaken too vigorously. The emulsion may be broken by the addition of small amounts of methanol.

Note 4   Yields of (*S*)-(+)-ethyl 3-hydroxybutanoate will be small. If several students combine and distill their products under reduced pressure, the suggested determinations will be carried out more easily.

Name   Section   Date

PART I   The Isomerization of Maleic Acid to Fumaric Acid

Reaction
equation

$$\underset{\text{Maleic acid}}{\begin{array}{c} H \quad\quad COOH \\ C \\ \| \\ C \\ H \quad\quad COOH \end{array}} \xrightarrow{\text{HCl}} \underset{\text{Fumaric acid}}{\begin{array}{c} H \quad\quad COOH \\ C \\ \| \\ C \\ HOOC \quad\quad H \end{array}}$$

Quantities   ____4.0 g____   _____

Mol. Wt.   _____   _____

Moles   _____   _____

Equivalents   _____   _____

Theoretical yield   _____ g

Actual yield   _____ g

Percentage yield   _____ %

PART II   Preparation of L-3-Phenyllactic Acid

Reaction
equation

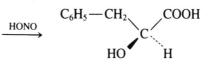

$$\underset{\text{L-Phenylalanine}}{\begin{array}{c} C_6H_5-CH_2 \quad COOH \\ C \\ H_2N \quad\quad H \end{array}} \xrightarrow{\text{HONO}} \underset{\text{L-3-Phenyllactic acid}}{\begin{array}{c} C_6H_5-CH_2 \quad COOH \\ C \\ HO \quad\quad H \end{array}}$$

Quantities   ____3.30 g____   _____

Mol. Wt.   _____   _____

Moles   _____   _____

Equivalents   _____   _____

Theoretical yield ＿＿＿＿＿ g

Actual yield   ＿＿＿＿＿ g

Percentage yield ＿＿＿＿＿ %

       mp   ＿＿＿＿＿＿

       $[\alpha]_D$ ＿＿＿＿＿＿

PART III   The Enzymatic Reduction of Ethyl Acetoacetate by Yeast

(a) Results of test with $FeCl_3$ solution:

(b) Observed rotation: $[\alpha]_D$ =

(c) Optical yield = ＿＿＿＿＿＿ %

**QUESTIONS
and
EXERCISES**

1. Name maleic and fumaric acids by the (E,Z)-system.

2. Why is maleic acid not only stronger than fumaric acid but also stronger than most simple aliphatic acids such as acetic acid?

3. Name L-phenylalanine and L-3-phenyllactic acid by the (R,S)-system.

4. The reaction of L-phenylalanine with aqueous nitrous acid proceeds with retention of configuration. The reaction is thought to involve the formation of an alkanediazonium ion followed by *two* successive S_N2 displacements, each with inversion of configuration, resulting in overall net retention. With this information to guide you, write a mechanism to explain the overall retention of configuration.

5. What product would you expect if L-phenylalanine were treated with sodium nitrite and concentrated hydrochloric acid? Explain.

6. L-3-Phenyllactic acid melts at 124–125°, whereas the racemic mixture, d,l-3-phenyllactic acid, melts at 98°. Explain. Where would you expect a mixture of 3 parts L- and 1 part D-enantiomer to melt? Explain.

7. What product would have resulted had we reduced ethyl acetoacetate with sodium borohydride, $NaBH_4$?

8. What structural feature in an organic compound gives a positive test with $FeCl_3$ solution?

9. Draw and name a pseudo three-dimensional structure for the enantiomer of (S)-$(+)$-ethyl 3-hydroxybutanoate.

*10. At which frequencies would strong absorptions appear in the ir spectrum of ethyl 3-hydroxybutanoate? Could absorption at this frequency also occur in the ir spectrum of ethyl acetoacetate? Explain.

*11. The following proton shift values (δ) appear in the nmr spectrum of ethyl 3-hydroxybutanoate:

Number of Protons	3	3	2	1	2	1
δ	1.15	1.28	2.35	3.15	4.05	4.15

Draw the structure of ethyl 3-hydroxybutanoate and identify the hydrogens that give the above values.

*12. In questions 10 and 11 why was it unnecessary to specify whether the spectra were those of the (S)-$(+)$, or the (R)-$(-)$-enantiomer?

Carbohydrates

For convenience the carbohydrates may be divided into three classes:

1. **Monosaccharides** (polyhydroxy aldehydes or ketones), which do not yield smaller units when hydrolyzed.

D(+)-glucose   D(−)-fructose

2. **Disaccharides,** which yield, when hydrolyzed, two monosaccharide units.

β-Maltose, a reducing disaccharide, yields two glucose units.

3. **Polysaccharides,** which, when hydrolyzed, yield many molecules of monosaccharides.

A small segment of the starch molecule ($n = 6,000–30,000$)

The following experiments illustrate some of the typical reactions of each class.

Precautions. *This experiment requires the use of α-naphthol; phenyl-hydrazine; hydrochloric, nitric, and sulfuric acids; iodine; acetic acid and acetic anhydride. See HAZARD CATEGORIES 2, 3, 4, 5, and 8, page 4.*

PROCEDURE

A. Preparation of Sugar Solutions

Prepare 4% solutions of the following sugars by dissolving 1 g of each in 25 mL of distilled water: glucose, fructose, sucrose, maltose, and lactose. Use these test solutions in Procedures B–D. Record all your observations on the report form and explain each test. Use equations whenever possible.

B. The Molisch Test

All substances having a carbohydrate grouping react with Molisch's reagent to give a purple color. The reaction is quite complex, but the test is a reliable indication of the presence of carbohydrates. Test one of the sugar solutions prepared in Procedure A as follows:

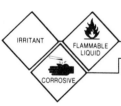

To 2 mL of the sugar solution in a test tube add 2 drops of Molisch's reagent (15% solution of α-naphthol in ethanol) and mix by shaking. Pour this solution slowly down the inside of a second test tube containing 2 mL of concentrated sulfuric acid. Hold the second test tube at an angle of approximately 30° while pouring so that two separate layers are formed without mixing. Observe any color change at the junction of the two liquids. Repeat the test using a suspension made by mixing a small micro-spatula measure of flour in 2 mL of water.

C. Benedict's or Fehling's Test for Reducing Sugars

All monosaccharides reduce Benedict's or Fehling's solutions whether they are aldehyde sugars (aldoses) or ketone sugars (ketoses). Disaccharides, on the other hand, are subdivided into two classes on the basis of their behavior toward these reagents. The two classes are (a) **reducing disaccharides,** which reduce Benedict's or Fehling's solutions, and (b) **nonreducing disaccharides,** which do not react with these solutions.

Test each of the sugar solutions prepared in Procedure A separately in the following manner:

Place 4 mL of Benedict's solution or 4 mL of freshly prepared Fehling's solution[1] in a test tube and heat the solution to gentle boiling. To the boiling solution add the sugar solution 2–3 drops at a time, heating the mixed solutions for at least one minute after each addition. Observe any color changes, and continue adding the sugar solution and heating until the blue color just disappears, but do not add more than 5 mL of any sugar solution. Note the appearance and color of any precipitate.

D. Osazone Formation

Monosaccharides and reducing disaccharides react with phenylhydrazine to yield **osazones.** These are condensation products that contain *two* phenylhydrazone groups rather than one, as you would expect from the reaction of a simple aldehyde or ketone. The osazones are crystalline derivatives useful in identifying sugars. Each osazone possesses a definite melting point and crystalline form, and the length of time required to form the osazone is a characteristic of the sugar being tested.

Support over a wire gauze or on an electric hot plate a 400-mL beaker half-filled with water, and heat the water to boiling. Label or mark 5 test tubes and to each add 4 mL of phenylhydrazine reagent (Experiment 14A) and 2 drops of saturated sodium bisulfite solution. To each test tube add 5 mL of the various sugar solutions prepared in Procedure A and label each test tube with the sugar it contains. Mix the solutions thoroughly and immerse all tubes at the same time in the beaker of boiling water. Record the time of immersion and the time at which each osazone begins to precipitate. (Note 1) Shake the tubes from time to time to prevent formation of a supersaturated solution of the osazone, and continue to heat the tubes for 20 minutes. Allow the tubes to cool slowly by removing them from the water bath and placing them in the test tube rack. Filter any crystals formed on the Hirsch filter and examine them. Allow the osazone of one sugar to dry and determine its melting point. If a low-power microscope is available, transfer a few osazone crystals to a microscope slide and examine them. Are they all alike? Sketch and name the osazone.

E. Hydrolysis of Disaccharides and Polysaccharides

(a) Inversion (Hydrolysis) of Sucrose

Place the remainder of the sucrose solution prepared in Procedure A in an Erlenmeyer flask, add 1 mL of dilute hydrochloric acid, and heat the mixture on the water bath for 30 minutes. Cool the solution and carefully neutralize it with 10% sodium hydroxide (use litmus). Test the neutralized solution with Benedict's or Fehling's solution as in Procedure C.

---

[1]Instructor's Guide.

(b) Starch (Acid Hydrolysis)

Place a beaker containing 120 mL of distilled water on a wire gauze or on an electric hot plate and heat the water to boiling. Mix 1 g of starch thoroughly with 10 mL of cold water, stirring and crushing until the suspension is free of lumps. Slowly stir the suspension into the boiling water and boil the mixture for 1–2 minutes after addition is complete. Test a portion of the starch solution with Benedict's reagent. Also test a small sample of the starch solution with a drop of iodine-potassium iodide solution.[1]

To 50 mL of the starch solution in a 125-mL Erlenmeyer flask add 5 drops of concentrated hydrochloric acid and heat the mixture for 30–40 minutes on the water bath. Cool the solution and carefully neutralize it with 10% sodium hydroxide. Test a portion of the cooled solution again with Benedict's solution and also with the iodine-potassium iodide solution. Result?

(c) Enzymes

To 50 mL of the starch solution in a 125-mL Erlenmeyer flask add about 2 mL of your own saliva. Mix and place in a warm (40°) water bath for 30 minutes. Again, test portions of the solution with Benedict's reagent and with iodine-potassium iodide solution. Result?

F. Acetylation of Cellulose

Cellulose (a polysaccharide) is also a polyhydric alcohol and may be esterified with acetic anhydride to give **cellulose acetate,** a compound in which two to three of the hydroxyl groups in each monosaccharide unit have been acetylated. Cellulose acetate is used commercially in the manufacture of such products as camera film, rayon, and adhesives.

In a 50-mL Erlenmeyer flask add 20 mL of glacial acetic acid, 7 mL of acetic anhydride, and 2 drops of concentrated sulfuric acid. Press 0.5 g of cotton beneath the surface of the liquid mixture with a stirring rod, making sure that the cotton is completely immersed. Allow the mixture to stand until the next laboratory period or heat it at 70–75° in a water bath for 30–45 minutes, stirring occasionally.

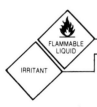

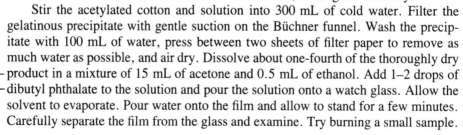

Stir the acetylated cotton and solution into 300 mL of cold water. Filter the gelatinous precipitate with gentle suction on the Büchner funnel. Wash the precipitate with 100 mL of water, press between two sheets of filter paper to remove as much water as possible, and air dry. Dissolve about one-fourth of the thoroughly dry product in a mixture of 15 mL of acetone and 0.5 mL of ethanol. Add 1–2 drops of dibutyl phthalate to the solution and pour the solution onto a watch glass. Allow the solvent to evaporate. Pour water onto the film and allow to stand for a few minutes. Carefully separate the film from the glass and examine. Try burning a small sample.

---

[1]Instructor's Guide.

Report your observations and attach a small sample of your product to your report sheet.

G.   Nitration of Cellulose

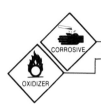

In a 100-mL beaker *cautiously* mix 10 mL of concentrated sulfuric acid and 10 mL of concentrated nitric acid. Heat the solution on the hot-water bath to 50° and then immerse a wad of absorbent cotton weighing about 0.5 g in the solution. Stir occasionally and allow the cotton to remain in this nitrating mixture no longer than 4 minutes. After removing the cotton, immediately immerse it in a large beaker of cold water. Change the water several times until a final rinsing no longer produces a wash water acid to litmus. Squeeze the cotton as dry as possible and spread it out on filter paper to dry. Cotton partially nitrated in this manner produces pyroxylin. Hold a piece of *dry* pyroxylin with your tongs in the flame of a burner. Result?

Dissolve a portion of the pyroxylin in 20 mL of a 50-50 mixture of alcohol and ether. Puddle and stir until completely dissolved. Decant onto a watch glass and allow the solvent to evaporate. (Note 2) Describe the appearance of the residual film. Lift the edge of the film and allow water to run beneath it to facilitate easy removal. Attach a sample to your report form.

Note 1   The osazones of lactose and maltose frequently form only upon cooling the reaction mixture. The partial hydrolysis of sucrose through long periods of heating results in the formation of sufficient glucose and fructose to give a positive osazone reaction. A positive reaction therefore is often erroneously reported for sucrose.

Note 2   Enough nitrated cotton should be dissolved so the mixture will pour onto the watch glass as a syrupy liquid.

Name   *Section*   *Date*

Carbohydrates

Results of the Molisch Test

Test with:

Glucose	Fructose	Sucrose	Maltose	Lactose	Starch

Results of the Benedict's or Fehling's Test

Test with:

Glucose	Fructose	Sucrose	Maltose	Lactose

Write an equation for the reaction that takes place when glucose is heated with Fehling's solution.

Results of the Osazone Test

Give time required for reaction of:

Glucose	Fructose	Sucrose	Maltose	Lactose

If osazone crystals were viewed through a microscope, make a sketch of the crystals. Name the osazone.

Color? _____

mp _____
(experimental)

mp _____
(literature)

Sketch of crystal

Results of Hydrolysis

(a) Inversion of surcrose. Describe your results with Benedict's or Fehling's solutions.

_____

_____

_____

(b) Acid hydrolysis of starch.

_____

Test with Benedict's reagent

Before hydrolysis _____   Positive or negative _____

After hydrolysis _____   Positive or negative _____

Test with iodine-potassium iodide

Before hydrolysis _____   Positive or negative _____

After hydrolysis _____   Positive or negative _____

Name _____   *Section* _____   *Date* _____

(c) Hydrolysis catalyzed by enzymes.

_____

   Test with iodine-potassium iodide   Positive or negative _____

   Test with Fehling's solution   Positive or negative _____

_____

Cellulose Acetate and Cellulose Nitrate

Describe the appearance of these films _____

_____

Describe the difference in burning properties of cellulose nitrate and cellulose acetate. _____

_____

(Attach samples here)

1. Why are the osazones of glucose and fructose identical?

2. Which other pairs of aldohexoses besides glucose and mannose give identical osazones?

3. Draw the structure for the disaccharide sucrose and point out why it is not a reducing sugar.

4. Why is the hydrolysis of sucrose referred to as "inversion"?

5. To what general class of compounds does cellulose acetate belong? Cellulose nitrate? Ethyl cellulose?

6. What is the maximum number of hydroxyl groups per glucose unit of cellulose that can be nitrated? If all these are nitrated, what commercial product is formed?

7. One of the earliest synthetic plastics, called *celluloid,* was manufactured from cellulose nitrate and camphor. Celluloid is now seldom seen. What obvious disadvantage keeps celluloid from being used in place of some of our modern plastics?

8. Draw the structure of β-D-glucose in such a way as to show the axial-equatorial distribution of the substituents on the ring. On the basis of your structure, suggest a reason for the observation that D-glucose is the most abundant monosaccharide in nature.

Proteins

Proteins are very complex, high molecular weight (10,000 to 10,000,000) structures that yield α-amino acids on hydrolysis.

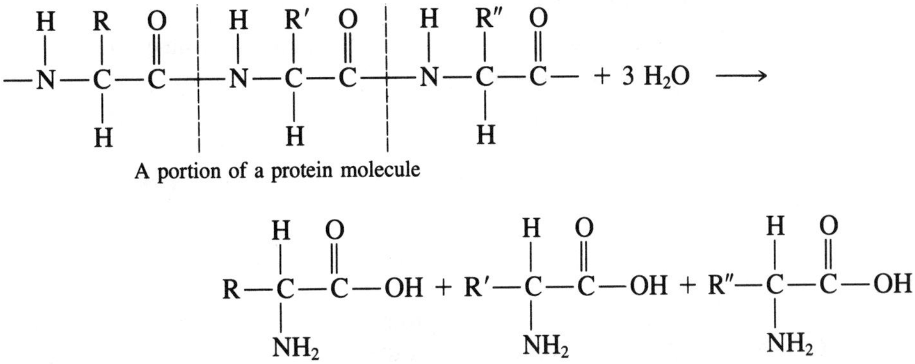

$$-\overset{\overset{\displaystyle H}{|}}{N}-\overset{\overset{\displaystyle R}{|}}{\underset{\underset{\displaystyle H}{|}}{C}}-\overset{\overset{\displaystyle O}{\|}}{C}\Big|-\overset{\overset{\displaystyle H}{|}}{N}-\overset{\overset{\displaystyle R'}{|}}{\underset{\underset{\displaystyle H}{|}}{C}}-\overset{\overset{\displaystyle O}{\|}}{C}\Big|-\overset{\overset{\displaystyle H}{|}}{N}-\overset{\overset{\displaystyle R''}{|}}{\underset{\underset{\displaystyle H}{|}}{C}}-\overset{\overset{\displaystyle O}{\|}}{C}- \ + \ 3\ H_2O \ \longrightarrow$$

A portion of a protein molecule

$$R-\overset{\overset{\displaystyle H}{|}}{\underset{\underset{\displaystyle NH_2}{|}}{C}}-\overset{\overset{\displaystyle O}{\|}}{C}-OH + R'-\overset{\overset{\displaystyle H}{|}}{\underset{\underset{\displaystyle NH_2}{|}}{C}}-\overset{\overset{\displaystyle O}{\|}}{C}-OH + R''-\overset{\overset{\displaystyle H}{|}}{\underset{\underset{\displaystyle NH_2}{|}}{C}}-\overset{\overset{\displaystyle O}{\|}}{C}-OH$$

α-Amino acids

The hydrolysis of a protein may be catalyzed by the action of enzymes, acids, or bases and may yield mixtures of as many as twenty or more α-amino acids.

Precautions. *This experiment requires the use of hydrochloric, nitric, and glacial acetic acids; soda lime; mercuric chloride; and phenol. See HAZARD CATEGORIES 2, 3, and 8, page 4.*

PROCEDURE

A. Hydrolysis of a Protein

In this part of the experiment you will hydrolyze either a sample of the phosphoprotein or casein from milk or a sample of the artificial sweetener aspartame (Nutra-Sweet), which is a dipeptide.

(a) Hydrolysis of Casein

Assemble a small-scale reflux apparatus as illustrated in Figures 0.1, 0.2, or 0.3 (page 25–29). Introduce into the boiling flask 20 mL of approximately 20% hydrochloric acid (Note 1), 0.5 g of casein, and a Boileezer or two. Heat the mixture under reflux for 30–45 minutes using a heating mantle, oil bath, or a very low flame. Control the heating very carefully until the casein is dissolved. While the mixture is refluxing, proceed with Procedures B, C, D, and E. At the end of the reflux period use test (a) below to determine whether hydrolysis is complete. If the test is *positive* add 3–5 mL of additional 20% hydrochloric acid and heat for an additional 15 minutes or until test 1 is *negative*. Cool the mixture, remove 5 mL of the hydrolysate for test 2 below, stopper, and save the remainder for Procedure H(a). (Note 2)

1. Carefully neutralize (using litmus) 1–2 mL of the hydrolysate with 10% sodium hydroxide solution. Do the amino acids precipitate? Make the solution definitely alkaline by adding an additional 1 mL of dilute sodium hydroxide. Add 4–5 drops of 2% copper sulfate solution. Do you obtain a positive biuret test? (See Procedure E.)

2. Amino acids, like other primary aliphatic amines, react with nitrous acid to liberate gaseous nitrogen.

$$R-\underset{\underset{NH_2}{|}}{\overset{\overset{H}{|}}{C}}-\overset{\overset{O}{||}}{C}-OH + HNO_2 \longrightarrow R-\underset{\underset{OH}{|}}{\overset{\overset{H}{|}}{C}}-\overset{\overset{O}{||}}{C}-OH + H_2O + N_2$$

An α-amino acid   An α-hydroxy acid

To 5 mL of your acidic protein hydrolysate slowly add an equivalent volume of 5% sodium nitrite. Repeat the test using a solution of 0.1 g of glycine in 5 mL of 10% hydrochloric acid. Record your observations.

(b) Hydrolysis of Aspartame

Assemble a small-scale reflux apparatus as illustrated in Figures 0.1, 0.2, or 0.3 (pages 25–29). Introduce into the boiling flask 20 mL of approximately 20% hydrochloric acid (Note 1), the contents of two 1-g packets of Equal (aspartame plus other ingredients), and a Boileezer or two. Heat the mixture under reflux for 45 minutes using a heating mantle, oil bath, or a very low flame. The solution will turn yellow, and a black precipitate may form. While the mixture is refluxing, proceed with Procedures B, C, D, and E. Allow the mixture to cool enough to handle. Add about 0.5 g of decolorizing charcoal, mix thoroughly, and filter by gravity into a small Erlenmeyer flask, using a minimum amount of distilled water to aid in the transfer. Remove 5 mL of the hydrolysate for the following test. Stopper and save the remainder for Procedure H(b).

Amino acids, like other primary aliphatic amines, react with nitrous acid to liberate gaseous nitrogen.

$$R-\underset{\underset{NH_2}{|}}{\overset{\overset{H}{|}}{C}}-\overset{\overset{O}{\|}}{C}-OH + HNO_2 \longrightarrow R-\underset{\underset{OH}{|}}{\overset{\overset{H}{|}}{C}}-\overset{\overset{O}{\|}}{C}-OH + H_2O + N_2$$

An α-amino acid   An α-hydroxy acid

To 5 mL of your acidic protein hydrolysate slowly add an equivalent volume of 5% sodium nitrite. Repeat the test using a solution of 0.1 g of glycine in 5 mL of 10% hydrochloric acid. Record your observations.

B.   Test for Nitrogen

Mix thoroughly, in a small evaporating dish, 0.5 g of casein or dried egg albumin with 1–2 g of finely powered soda lime (sodium hydroxide and calcium oxide). Transfer the mixture to a dry test tube and heat over a low flame. Note cautiously the odor of the gas that evolves and test it with a piece of moist red litmus held over the mouth of the test tube. What was the product formed?

C.   Test for Sulfur

To about 0.5 g of dried egg albumin in a 50-mL Erlenmeyer flask add 10–15 mL of 10% sodium hydroxide and very gently boil for about 15 minutes. The mixture will have a tendency to froth easily. Cool the solution and make it acidic with hydrochloric acid. Bring it to a boil again, after having placed a piece of moist lead acetate paper over the mouth of the flask. Observe the change in the acetate paper. What material is formed? What amino acids give a positive test?

D.   Xanthoproteic Test

Proteins that have α-amino acid units containing aromatic nuclei may be nitrated to give yellow compounds, which deepen in color when treated with alkali. You have unwittingly performed this test if you ever have spilled nitric acid on your fingers.

To 3 mL of egg-white solution (Note 3) in a test tube add 10–15 drops of concentrated nitric acid and warm gently. Observe and record any color change. Cool the solution and neutralize it with 10% sodium hydroxide. Note and record any change in color.

E.   The Biuret Test

The biuret reaction is a general test for proteins and for compounds that have in their structures certain multiple amide linkages. Perhaps the simplest compound to give

a positive test is biuret itself,

$$\underset{\substack{}}{H_2N-\overset{\displaystyle\overset{O}{\|}}{C}-\overset{\displaystyle\overset{H}{|}}{N}-\overset{\displaystyle\overset{O}{\|}}{C}-NH_2}$$

A positive biuret test is indicated by the production of a violet-pink color when alkaline solutions of compounds containing the above structural units are treated with a dilute solution of copper sulfate.

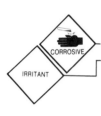

In each of two dry test tubes place 1 g of urea. Heat one test tube gently until the urea melts. Cautiously note the odor of the gas that evolves and test it with a piece of moist red litmus held over the mouth of the test tube. Continue heating gently until the material shows a tendency to solidify. Dissolve the white residue in 3–4 mL of warm water and filter. To the filtrate add an equal volume of 10% sodium hydroxide, and mix. To this alkaline solution add 1–2 drops of 2% copper sulfate. Shake the mixture and observe the color. Record your results and write the equation for the formation of biuret.

Without heating the other sample of urea, add 3–4 mL of water, make alkaline, and add copper sulfate as before. Result?

F. Amphoteric Nature of Proteins

Proteins, like the amino acids, are amphoteric substances, that is, they show both acidic and basic properties. Depending on the pH of the solution, a protein will exist as mixtures of positively charged, negatively charged, and neutral species that we will represent schematically with the following forms.

$$\underset{I}{H_3N^+-(\text{protein})-CO_2H} \qquad \underset{II}{H_3N^+-(\text{protein})-CO_2^-} \qquad \underset{III}{H_2N-(\text{protein})-CO_2^-}$$

The actual numbers of free amino and carboxyl groups will not be just one of each as shown but will vary according to the composition of the protein. Depending in part on the relative numbers of free amino and free carboxyl groups and on the pH, a protein in solution may behave as an acid (represented by Form I), as a base (represented by Form III), or a neutral species (represented by the dipolar Form II). If just the right amount of acid or base is added to a protein solution, the numbers of positively and negatively charged centers can be made equal, and the protein will become electrically neutral (essentially all in Form II). The pH at which electrical neutrality is achieved is called the **isoelectric point.** At their isoelectric points the solubility of proteins is at a minimum, and some, such as casein and metaprotein, are essentially insoluble.

In this part of the experiment you will prepare metaprotein by the partial hydrolysis of egg albumin and examine its amphoteric properties.

Into each of two test tubes place 5 mL of egg-white solution. To one test tube add 2 mL of 10% sodium hydroxide and to the other add 2 mL of 10% hydrochloric acid. Warm the contents of both test tubes on the water bath for about 30 minutes.

Cool, then neutralize the acid solution (use litmus) by adding dropwise 10% sodium hydroxide, and neutralize the alkaline solution by adding dropwise 10% hydrochloric acid (use litmus). What occurs in each case at the neutral point? Remove the precipitate from each test tube by filtration and test in the following manner: (1) Transfer a small portion of the precipitate to a test tube, add 2–3 mL of 10% hydrochloric acid, and shake well. Result? (2) Place a small portion of the precipitate into a second test tube and add 2–3 mL of 10% sodium hydroxide solution. Shake well and again observe your result. (3) In a third test tube shake a small portion of the precipitate with 2–3 mL of distilled water. Result?

G. Precipitation with Salts of Heavy Metals

Into each of four test tubes place 2–3 mL of the egg-white solution. To one test tube slowly add 10% mercuric chloride solution. Observe the result. Repeat the test by slowly adding 10% solutions of ferric chloride, lead acetate, and copper sulfate. Record your results.

H. The Separation of α-Amino Acids by Paper Chromatography

α-Amino acids may be separated by either paper or thin layer chromatography as well as by other more advanced techniques. In this part of the experiment you will separate the amino acids in the hydrolysate from Part A, using either of two well-established solvent systems, 80:20 phenol:water or 60:15:25 1-butanol:acetic acid:water. The phenol-water system is described in Part H(a) for the separation of the casein hydrolysate and the 1-butanol-acetic acid-water system is described in Part H(b) for the aspartame hydrolysate; however, either solvent system may be used with either hydrolysate. Your instructor will tell you which solvent system to use.

(a) Separation of Casein Hydrolysate Using Phenol-Water

The technique used for the separation of α-amino acids is similar to that described in Part II, Experiment 7, except your technique must be somewhat more refined. (Note 3)

An approximately 10 × 16-cm sheet of the type of Whatman chromatography paper previously described may be used. Take special care to handle the paper only by its upper end. Place the paper on a clean sheet of notebook paper, draw a *pencil* line as before, but make five marks at 1.5-cm intervals. Identify these marks as A, P, T, M, and H. Using a separate micropipet for each sample, place at M a small (2.0-mm) spot of a mixture of known amino acids made by mixing equal volumes of 0.1M solutions of aspartic acid, phenylalanine, and tyrosine. (Note 4) At A, P, and T place spots of aspartic acid, phenylalanine, and tyrosine solutions, respectively. At H place a small spot of the hydrolysate saved from Procedure A(a). Allow the spots to dry and then spot them a second time to make certain that sufficient material will be present for any easily observed resolution. When the spots are dry, position the paper in the developing jar as before and secure it in position. (Note 5) Through

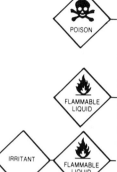

a long-stemmed funnel or with a pipet and pipet bulb, transfer to the bottom of the developing jar 15–20 mL of a *freshly prepared* 80% phenol solution. Make certain that the level of the phenol solution does not rise up to the level of the spots on the paper. Close the mouth of the developing jar with aluminum foil and allow the solvent to rise up the paper for approximately 90 minutes. At the end of this time remove the paper and mark the upper limit of the solvent front. Wash off the excess phenol by rinsing the paper thoroughly on both sides with acetone from a squeeze bottle. Catch the wash acetone in the developing jar and do not splash it on yourself or on the desk. After the acetone has evaporated (**Hood!**), spray the paper lightly with a 2% solution of ninhydrin in 95% ethyl alcohol. (Note 6) The paper must be thoroughly sprayed with ninhydrin but should not be dripping wet. Hang the paper up to dry. Color spots usually begin to appear within 5–15 minutes; however, color development is faster and more reliable if the nearly dry paper is heated in the oven at 105° or under a heat lamp for 10–20 minutes. Circle each colored spot and measure the distance from its center to the sampling line. Calculate and record all Rf values given in Table 20.1.

Unless your instructor directs otherwise, pour the acetone rinse from your developing jar into the container provided. Rinse the jar with a little acetone, collecting the rinse solvent in the container. Wash the developing jar thoroughly with water, then with soap and water, and finally with distilled water.

(b) Separation of Aspartame Hydrolysate Using 1-Butanol-Acetic Acid-Water

The technique used for the separation of α-amino acids is similar to that described in Part II, Experiment 7, except your technique must be somewhat more refined. (Note 3)

A 10 × 16 sheet of Whatman chromatography paper of the type previously described may be used. (Note 7) Take special care to handle the paper only by its upper end. Place the paper on a clean sheet of notebook paper, draw a *pencil* line as before, but make four marks at 1.5-cm intervals. (Note 7) Identify these marks as A, P, M, and H. Using a separate micropipet for each sample, place at M a small (2.0-mm) spot of a mixture of known amino acids made by mixing equal volumes of 0.1M solutions of aspartic acid and phenylalanine. At A and P place spots of aspartic acid and phenylalanine solutions, respectively. At H place a small spot of the hydrolysate saved from Procedure A(b). Allow the spots to dry. Spot H should be the only one that requires respotting; respot it a second and third time to make certain that sufficient material will be present for an easily observed resolution. When the spots are dry, position the paper in the developing jar as before and secure it in position. (Note 5) Through a long-stemmed funnel, transfer to the bottom of the developing jar 20 mL of a mixture made from 12 mL of 1-butanol, 3 mL of glacial acetic acid, and 15 mL of distilled water in a small Erlenmeyer flask. Make certain that the level of the solvent mixture does not rise up to the level of the spots on the paper. Close the mouth of the developing jar with aluminum foil and allow the solvent to rise up the paper for approximately 1.5–2 hours. At the end of this time

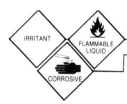

Table 20.1 *Approximate Percentage Composition and Rf Values of Amino Acids in Casein*

Amino acid	Formula	%	Rf value[a]	Rf value[b]
Cystine	S—CH$_2$CH(NH$_2$)COOH | S—CH$_2$CH(NH$_2$)COOH	0.35	0.16	0.09
Aspartic acid	HOOC—CH$_2$CH(NH$_2$)COOH	7.1	0.32	0.23
Glutamic acid	HOOC—CH$_2$CH$_2$CH(NH$_2$)COOH	23.3	0.40	0.28
Glycine	H$_2$N—CH$_2$COOH	2.7	0.42	0.23
Serine	HOCH$_2$CH(NH$_2$)COOH	7.7	0.43	0.22
Threonine	CH$_3$CH(OH)CH(NH$_2$)COOH	4.9	0.51	0.26
Alanine	CH$_3$CH(NH$_2$)COOH	3.0	0.59	0.30
Tyrosine	HO—⟨benzene ring⟩—CH$_2$CH(NH$_2$)COOH	6.3	0.62	0.45
Lysine	H$_2$N(CH$_2$)$_4$CH(NH$_2$)COOH	8.2	0.71	0.12
Valine	(CH$_3$)$_2$CHCH(NH$_2$)COOH	7.2	0.75	0.51
Arginine	H$_2$N—C(=NH)—NH(CH$_2$)$_3$CH(NH$_2$)COOH	4.1	0.76	0.15
Methionine	CH$_3$S(CH$_2$)$_2$CH(NH$_2$)COOH	3.4	0.77	0.50
Leucine	(CH$_3$)$_2$CHCH$_2$CH(NH$_2$)COOH	9.2	0.79	0.70
Phenylalanine	⟨benzene ring⟩—CH$_2$CH(NH$_2$)COOH	5.0	0.82	0.60
Proline	⟨pyrrolidine ring with H, COOH, N—H⟩	11.3	0.85	0.34

[a] 80% Phenol
[b] 60:15:25 1-Butanol:Acetic acid:Water

remove the paper and mark the upper limit of the solvent front. Allow the paper to dry standing on a watch glass in the hood; then spray the paper lightly with a 2% solution of ninhydrin in 95% ethyl alcohol. (Note 6) The paper must be thoroughly sprayed with ninhydrin but should be not be dripping wet. Hang the paper up to dry. Colored spots usually begin to appear within 5–15 minutes; however, color development is faster and more reliable if the nearly dry paper is heated in the oven at 105°, over (not on) a hot plate, or under a heat lamp for 5–10 minutes. Circle each colored spot and measure the distance from its center to the sampling line. Calculate and record all Rf values for each amino acid on your report form. Compare your Rf values with those given in Table 20.1

I. Detection of Aspartame in a Diet Drink

This procedure is exactly the same as that described in Experiment 7, Part III, page 96; however, in this variation you will be trying to detect the presence of the artificial sweetener aspartame in a diet drink and will require a different developing solvent and a different brand of coated TLC film. Obtain a diet drink whose label indicates the presence of aspartame or NutraSweet. Pour a small amount into a small beaker to allow the carbon dioxide to escape; otherwise, your capillary pipets will not work well. Dissolve 1 packet of Equal (or about 15–20 mg of authentic aspartame) in 15 mL of distilled water in a 25-mL Erlenmeyer flask. Make two 1.5-2 mm spots on 25 × 100-mm plates prepared from Aldrich Z-12,277-7 or Z-12,278-5 coated TLC film (Eastman TLC film does not give satisfactory results), one from the diet drink and the other from the aspartame "standard." The spot from the diet drink should be dried and respotted for a total of at least three applications. The developing solvent is prepared by mixing 3 mL of methylene chloride, 2 mL of methanol (***Caution!*** *Liquid and vapors are toxic.*), and 1 mL of glacial acetic acid (***Caution!*** *Liquid and vapors are corrosive.*) in the 400-mL beaker. Allow the plate to develop until the solvent front has ascended to within 1.5 cm of the top of the plate. Remove the plate and allow it to dry. Wearing disposable plastic gloves and working in the exhaust hood, spray the plate with the ninhydrin reagent. After the plate has dried, place it in the oven at 100°–110° for 5–10 minutes or warm it gently on a hot plate turned to a low setting. The presence of aspartame will be indicated by an orangish red to rust colored, elongated oval spot with an Rf value of 0.5–0.6. Attach the plate to your report form.

Note 1   Constant boiling hydrochloric acid has a composition of slightly over 20% (depending on the pressure of the distillation) and is a common standard acid. For this experiment approximately 20% hydrochloric acid may be prepared by mixing equal volumes of concentrated hydrochloric acid and distilled water.

Note 2   The hydrolysate should be clear to pale yellow. If it is cloudy or darkly colored, add 0.5 g of decolorizing charcoal while the solution is still warm. Mix thoroughly and filter by gravity into a 50-mL Erlenmeyer flask.

Note 3   Fingerprints will show up as colored spots when the chromatogram is developed if the paper is handled carelessly. Wear plastic gloves of the throwaway type during the preparation of the chromatogram and the spraying operation to (1) prevent you from coming into contact with the solvent system, (2) protect your chromatogram from contamination from your fingers, and (3) prevent your hands from being discolored by the ninhydrin reagent.

Note 4 For best comparison with the hydrolysate, the standard amino acid solutions should be made acidic with the approximately 20% hydrochloric acid. Add about 1 mL of the hydrochloric acid for every 20 mL of amino acid solution.

Note 5 It is very important that the spots be dry before solvent moves over them; otherwise they will become diffuse.

Note 6 Easy-to-use aerosol sprays are commercially available for α-amino acid chromatography (see Instructors Guide), however, a Windex glass cleaner bottle or any small sprayer may be used.

Note 7 If you are using a wide-mouthed, 1-quart Mason jar for a developing chamber, you may use a 6.66-cm strip of chromatography paper (one-third of a 10 × 10-cm sheet). Mark the upper end of the paper for the wire support as before. Place the four spots at 1.2-cm intervals. After the spots are dry, staple the paper onto the wire support and make any adjustments necessary for the paper to fit the jar. Remove the paper, add the solvent mixture, and swirl it about the jar. Carefully place the paper in the jar, hooking the wire support in position. Cover the mouth of the jar with aluminum foil.

REPORT
20

Name ____________ *Section* ____________ *Date* ____________

Proteins

Hydrolysis of a Protein

(a) Result of the biuret test: ____________

(b) Reaction with nitrous acid: ____________

(c) Reaction of glycine with nitrous acid: ____________

Test for Nitrogen

Nature of the product formed ____________

Complete the following equation.

$$R-\overset{\overset{\displaystyle O}{\|}}{C}-NH_2 + NaOH \longrightarrow$$

Test for Sulfur

What gas evolved when the acidified albumin hydrolysate was heated?

____

Write the equation for its reaction with lead acetate, $\left(CH_3-\overset{\overset{\displaystyle O}{\|}}{C}-O\right)_2 Pb$.

Xanthoproteic Test

Describe the color change which occurred when a protein sample was treated with nitric acid. ____________

Describe the color change after addition of sodium hydroxide. ____________

____

____

The Biuret Test

Describe your results when a heated sample of urea was tested for biuret. _____

_____

What was the result when the unheated sample of urea was tested? _____

_____

Write the chemical equation for the preparation of biuret from urea.

Name _____   Section _____   Date _____

Amphoteric Nature of Proteins

Sample	Result when neutralized
Egg white + 10% HCl	
Egg white + 10% NaOH	

What was the result when precipitated egg white was treated with acid? _____

_____

When treated with alkali? _____

Action of Heavy Metal Salts on Proteins

Metal salt	Action on egg white
$FeCl_3$	
$\left(CH_3 - \overset{\overset{\textstyle O}{\|}}{C} - O \right)_2 Pb$	
$HgCl_2$	
$CuSO_4$	

The Separation and Identification of α-Amino Acids (*Rf* Values)

Amino acid	Experimental	Literature
Aspartic acid	_____	_____
Phenylalanine	_____	_____
Tyrosine	_____	_____

(Attach chromatogram here.)

Name   Section   Date

The Detection of Aspartame in a Diet Drink

Describe your observations and attach TLC plate here.

1. Why did the sample of casein hydrolysate give a continuous streak rather than separate spots?

2. What explanation can you offer for the fact that on the casein hydrolysate chromatogram denser areas occur at Rf values of 0.30–0.45 and again at 0.80–0.85? (*Hint:* Consult Table 20.1.)

3. Write the structures of three sulfur-containing α-amino acids. Name each.

4. Write the structures of two α-amino acids that would give a color with nitric acid. Name each.

5. What was the gas that evolved from urea when the sample was heated?

6. Which amino acids in a protein hydrolysate are responsible for the black color in the lead acetate test? What is the black color?

7. Why is egg white frequently suggested as an antidote for certain types of heavy metal poisoning?

8. What is a peptide? Write the structure for glycylglycylalanine. To what classification does it belong?

9. Assign a name to the following structure:

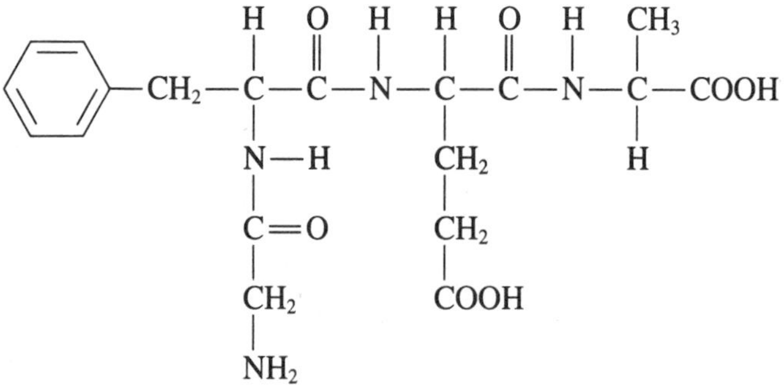

10. What protein readily available in every grocery store could have served for the preparation of a hydrolysate in Part A?

11. Which of the α-amino acids listed in Table 20.1 would have made up a considerable percentage of the total had we used a hydrolysate prepared from wool?

A Biosynthesis

PART I   The Preparation of Ethanol

A biochemical synthesis of ethanol, practiced since antiquity and of great commercial importance today, is that achieved through the fermentation of starches and sugars. Fermentation is a very complex, multistep biochemical process in which relatively large structures such as starches are broken down into simple sugars through the catalytic action of enzymes—specific enzymes being required for each step of the degradation. Every brewery, winery, and bakery is dependent upon the chemical changes brought about by fermentation. Yeast is used in the baking process to provide the enzymes (Greek *en*, in; *zyme*, leaven) that initiate fermentation for the primary purpose of providing the leavening action created by the evolution of carbon dioxide. Volatile organic compounds, of course, are removed during the baking process. When a yeast fermentation is carried out for nonbaking purposes, the result is a final potpourri that includes, in addition to ethanol, some aldehydes, ketones, acids, vitamins, minerals, and numerous other components.

It is beyond the scope of this manual to outline here the numerous interacting biochemical reactions involved in the enzymatic conversion of carbohydrates into the various organic substances cited above. For our purpose we shall simply indicate the fermentation process as a chemical reaction that stoichiometrically, at least, takes place according to the following equation:

$$C_6H_{12}O_6 \xrightarrow{\text{zymase}} 2\ C_2H_5OH + 2\ CO_2$$

As starting material we shall use a simple hexose such as glucose (grape sugar) or fructose (fruit sugar). The sugar in any naturally sweet substance may be fermented, but for our preparation we will begin with freshly prepared apple cider. No yeast is needed because the required enzymes are already present.

According to the preceding equation the fermentation reaction that leads to ethanol requires no oxygen but does produce carbon dioxide. Therefore, the reaction must be carried out under anaerobic conditions in a partially closed system with provision made for allowing carbon dioxide to escape but no air to enter. Without this precaution, acetic acid (vinegar) would be our end product, as we shall see in Part II of the experiment. Indeed, if oxygen were freely available only carbon dioxide and water would be produced.

PROCEDURE   Fill a 250-mL Erlenmeyer flask or a pop bottle to within 2 inches of the top with freshly prepared apple cider. (Note 1) Stopper the flask with a one-hole rubber stopper through which a section of bent glass tubing may be inserted and led into a CO_2 trap as illustrated in Figure 21.1. Set the cider in a warm place (25°) and allow fermentation to proceed for 2–3 weeks or until there is no further evolution of carbon dioxide. Filter the product to remove some of the sediment that accumulates during the fermentation, then carry out the following experiments.

A. GLC Analysis of Ethanol

Prepare a fractional distillation assembly (Fig. 4.1) using a 500-mL round-bottomed flask and place in it the entire 250 mL of hard cider filtrate. Add a Boileezer and, using a Bunsen burner, heating mantle, or oil bath (Note 2) as your heat source, collect 35–50 mL of distillate in a 100-mL receiving flask. Discard the residual liquid in the boiling flask and replace the latter with the 100-mL flask containing the 35–50 mL of distillate just collected. Using the same fractionating column, redistill and collect only distillate that can be brought over under a temperature of 80° (approximately 10–15 mL).

FLAMMABLE LIQUID

Using the gas chromatograph, determine the concentration of ethanol in the fraction that you have collected. You may follow the general procedure described in Experiment 7-IV. A column packed with Porapak-S and maintained at a temperature of 150° and a flow rate of 60 mL/second will be satisfactory. Use 1-μL samples and adjust the recorder attenuation to give peaks of maximum height. Also analyze and record on the same chromatogram a sample of absolute (200 proof) ethanol from the supply room. Compare the chromatograms and, using the supply-room sample as a standard, calculate the percentage composition of the fraction you have collected.

Figure 21.1   Assembly for carrying out a fermentation on a small scale. (a) Fermenter: (b) carbon dioxide trap.

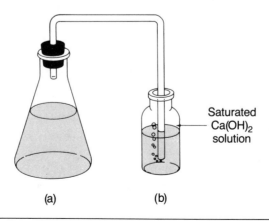

Saturated Ca(OH)$_2$ solution

(a)   (b)

Record your calculations on the report form and attach chromatograms. Your chromatogram should resemble that shown in Figure 21.2.

B.   Oxidation Test

Follow the procedure outlined under Experiment 12 (Part I-D) and carry out the oxidation test on a sample of cider alcohol. Result?

C.   Iodoform Test

Follow the procedure outlined under Experiment 12 (Part I-E) and carry out the iodoform test on a sample of cider alcohol. Result?

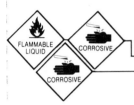

D.   Esterification

In a test tube place 2 mL of glacial acetic acid and 1 mL of your alcohol sample. Add one drop of concentrated sulfuric acid. (**Caution!**   *Use a Pasteur dropper!*) Mix well

---

Figure 21.2   GLC chromatograms of fermented cider distillate and 200 proof ethanol. Point of injection of samples indicated by (↑).

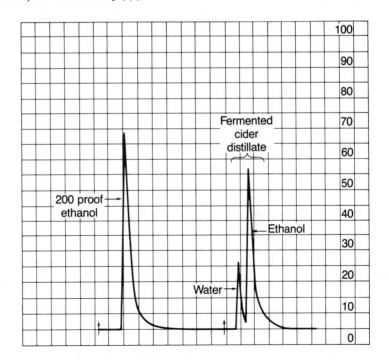

---

and place in a boiling water bath for 15 minutes. Dilute with 2–3 mL of water, pour onto a watch glass or into a Petri dish, and, with a wafting motion of the hand over the mixture, determine if you can detect the fruity odor of ethyl acetate. If you are not familiar with the odor of ethyl acetate, your instructor will provide a sample. Record all results of Tests B–D on the report form.

Note 1   Freshly prepared apple cider that is unpasteurized and that contains no preservatives must be used for this experiment. Frozen apple or grape juice concentrate containing no preservatives or additives (for example, vitamin C) may be used successfully if fresh cider is not available.

Note 2   See discussion of heating methods for distillation in Experiment 3.

PART II   The Preparation of Acetic Acid

The biosynthesis of acetic acid, unlike the preparation of ethyl alcohol, is an enzyme-catalyzed *oxidation* reaction that takes place when fruit juices are fermented in the presence of air. The alcohol that is produced first when apple cider ferments is further acted upon by various species of *Acetobacter* when the fermentation process is carried out in an open vessel. The reaction produces a vinegar in which acetic acid is the principal component and may be written in the following equation form:

$$CH_3CH_2OH + O_2(air) \xrightarrow{Acetobacter} CH_3COOH + H_2O$$

The preparation of vinegar for use as a condiment and preservative is another chemical reaction little understood but practiced by humans for thousands of years. A number of fruit juices may be used as starting material for the preparation of vinegar, but that obtained from apples is the starting material for brown or cider vinegar—the kind we will prepare in the present experiment.

PROCEDURE   Place 250 mL of freshly prepared apple cider in a 500-mL beaker or a glass jar, cover with a wire gauze, and set in a warm (25°) place. (Note 1) Allow the cider to ferment for 2–3 weeks, periodically stirring it to break up the slimy film ("mother of vinegar") that forms on the surface and prevents free contact with air. When fermentation is complete, filter the mixture to remove the scum and residue that accumulate, then carry out the following experiments on the relatively clear filtrate.

Test 1   Determine the acidity (expressed as molarity of acetic acid) of a sample of the vinegar you have prepared by diluting a 25-mL portion with 50 mL of distilled water and titrating with a standardized sodium hydroxide solution. A base strength approximately $0.5M$ is needed. Use phenolphthalein as an indicator. (Note 2) The concentration of acetic acid in commercial vinegar usually is fixed at 5% (approximately $0.83M$). Next, take a pH measurement on a sample of the vinegar.

Test 2 Distill a 50-mL sample of your vinegar using a Bunsen burner, heating mantle, or oil bath (Note 3) as your heat source. Distill until only 1–2 mL remain in the boiling flask. (Note 4) Distillation will separate the volatile acetic acid from any solid acids that may be in solution. In addition to acetic acid, cider vinegar usually contains certain amounts of malic and tartaric acids. Again titrate a sample of the distillate as was done in Test 1, except that the dilution with water is unnecessary. Also, measure the pH of the distillate.

Test 3 Analyze a sample of commercial white vinegar by determining its molarity and pH values.

Record the results of your measurements in Parts (a–c) on your report form.

Test 4 From concentration measurements and pH values obtained in Tests 2 and 3, determine a value for the pK_a of acetic acid.

You may find the following relationships useful.

$$K_a = \frac{\text{Concentration of hydronium ion} \times \text{Concentration of acetate ion}}{\text{Concentration of undissociated acid}}$$

$$K_a = \frac{[H_3O^+][CH_3COO^-]}{[CH_3COOH]}$$

$$[H_3O^+] = [CH_3COO^-]$$

$$pH = \log\frac{1}{[H_3O^+]} = -\log[H_3O^+]$$

$$pK_a = -\log K_a$$

Note 1 See Note 1 under Part I.

Note 2 The yellow color of undiluted vinegar will mask the phenolphthalein color change at the end point.

Note 3 See discussion of heating methods for distillation in Experiment 3.

Note 4 Frothing may present a problem. In this case, carry out the distillation as judiciously and for as long as possible.

Name   *Section*   *Date*

Ethanol

GLC Analysis of ethanol

Peak areas of prepared ethanol: C_2H_5OH _____; H_2O _____

Sample contained _____% ethanol; _____% water.
(Attach chromatogram)

Oxidation of Ethanol

Explain the color changes that occurred when a sample of ethanol was treated with

chromic acid. _____

_____

Write a *balanced* equation for the reaction that took place by completing the
following:

$$C_2H_5OH + \quad K_2Cr_2O + \quad H_2SO_4 \longrightarrow \quad CH_3COOH + \quad K_2SO_4 + \quad Cr_2(SO_4)_3 + \quad H_2O$$

Iodoform Test

Describe the results of the iodoform test. _____

_____

Write a *balanced* equation for the reaction that took place by completing the
following:

$$C_2H_5OH + \quad NaOH + \quad I_2 \longrightarrow \quad CHI_3 + \quad HCOO^-Na^+ + \quad NaI + \quad H_2O$$

Esterification Test

Describe the results of the esterification test. _____

_____

Acetic Acid in Vinegar

	Apple cider vinegar			Commercial white vinegar		
	pH	[CH$_3$COOH]	pK_a	pH	[CH$_3$COOH]	pK_a
Average Values	2.70 2.95	0.823 0.727	5.06 5.76	2.50 2.48	0.850 0.839	4.13 4.88
Experimental Values						

K_a for acetic acid $= 1.78 \times 10^{-5}$; p$K_a = 4.75$.

1. Champagne bottles are made of heavy glass and stoppers are wired tight. What is the reason for this?

2. Some beers are bottled before fermentation is complete, while others are bottled after fermentation is complete but are carbonated and bottled while very cold. Explain.

3. A natural wine usually has an alcoholic content no greater than 12–13% by volume. Why can the alcoholic content not exceed this?

4. King Arthur and his knights on occasion probably spent some time at the Round Table drinking mead. What is mead?

5. If a fractional distillation had been carried out a third or even a fourth time on our sample of fermented cider, could we have obtained pure ethanol? Explain.

6. Consult your text and describe the Weizmann fermentation. What products resulted from this historically and commercially important achievement?

7. What are the starting materials and methods for the production of ethanol other than by fermentation of carbohydrates? Write an equation to illustrate.

8. Consult your text or some other literature source to learn the meaning of *pyro-ligneous acid*. From what is it made? What is its composition?

Preparation of Cyclohexene

The dehydration of alcohols with acids is a general laboratory method for preparing alkenes. Strong mineral acids, such as sulfuric or phosphoric acids, usually are used as catalysts. In this experiment cyclohexanol is dehydrated to produce cyclohexene according to the following equation.

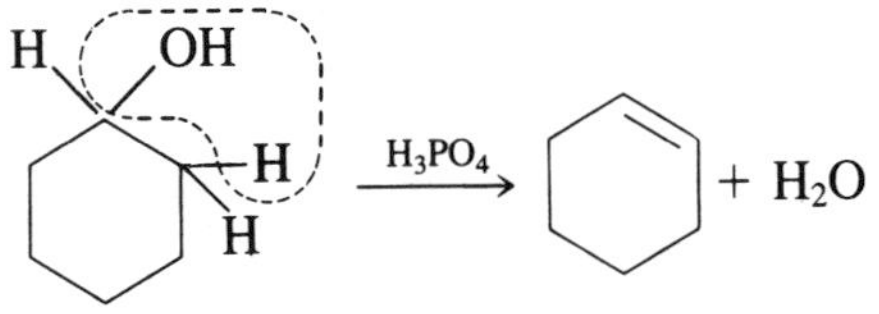

Cyclohexanol   Cyclohexene

Cyclohexene (bp 83°), along with some water, is distilled from the reaction mixture as it is produced by careful heating of the mixture at or near 100°. Unreacted cyclohexanol (bp 161°) remains in the boiling flask to be further acted upon by the acid. By careful control of the temperature the reaction can be made to go to completion. The small amount of mineral acid that invariably appears along with the product is neutralized by washing the product with sodium carbonate. Traces of water are removed from the crude cyclohexene by drying the liquid over anhydrous calcium chloride (a salt that forms hydrates and removes water from many organic liquids).

Precautions.   *This experiment requires the use of cyclohexanol; sulfuric or phosphoric acid. See HAZARD CATEGORIES 1, and 2, page 4.*

PROCEDURE

Set up a distillation apparatus as illustrated in Figure 3.1, using a 100-mL round-bottomed flask. Either a Bunsen burner or an electric heating mantle may be used, but, in this experiment, better control of heating is possible with the burner. If you use a burner, support the flask on a ceramic-fiber centered wire gauze. If you use a mantle, support it in such a way that it can be removed quickly if necessary to control the temperature of the reaction mixture. (Note 1) Inasmuch as the product is a very volatile and flammable hydrocarbon, use a special vented receiver of the type

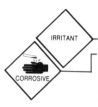

illustrated in Figure 22.1, and make certain that the ground-glass joints are properly sealed. If an electric heating mantle is your heat source, the hose leading from the side-arm of the adapter may be eliminated. Cool the receiver in crushed ice and water. Place 20 g (21 mL, 0.20 mole) of cyclohexanol (sp. gr. 0.96) and 5.0 mL of 85% phosphoric acid (Note 2) in the distilling flask (use a long-stemmed funnel) and mix thoroughly. Add a Boileezer or two to the mixture, start the water circulating gently through the condenser, and carefully heat the reaction mixture. If a burner is your heat source, use a small flame. Control the heating operation by holding the burner in your hand, moving the burner as necessary to maintain the temperature. The temperature of the vapor condensing on the thermometer bulb should never register in excess of 100°. If a mantle is your heat source, you will have to adjust the voltage carefully to control the temperature, removing the mantle temporarily if necessary to keep the vapor temperature below 100°. Continue the distillation until only 5–6 mL of residue remains in the boiling flask. Saturate the aqueous layer of the distillate with solid sodium chloride (approximately 0.5 g) and add enough 10% sodium carbonate solution to make the aqueous layer basic to litmus. Transfer the mixture to a separatory funnel and separate the lower aqueous layer from the hydro-carbon. Pour the cyclohexene through the neck of the funnel into a Erlenmeyer flask and add 1–2 g of anhydrous calcium chloride. During the drying interval (Note 3), disassemble the cooled distillation appararatus and clean it cautiously as follows. Rinse the condenser and adapter with water to remove traces of acid. Add *cold* water to the distilling flask, carefully discard the contents of the flask in the sink in the hood (taking care not to wash the Boileezers into the sink), and rinse the flask thoroughly with water. Dry the condenser and adapter and reassemble your distillation apparatus, but this time use a distilling flask of 50-mL capacity. Transfer the cyclohexene

Figure 22.1 Receivers for highly volatile and flammable distillates vented via rubber tubing over side of laboratory desk.

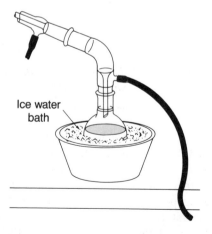

Ice water
bath

to the distilling flask by pouring it through a clean, dry, long-stemmed funnel fitted with a small cotton plug to retain the calcium chloride. (Note 4) Add a Boileezer or two and distill, collecting the principal portion within a boiling range of 80–85°. Yield, 10–12 g (79–80%). After weighing your product and calculating your yield, (Note 5) use small samples of your product to perform the bromine and Baeyer tests described in Experiment 8, Part I-B. Save the remainder of your cyclohexene for use in Experiment 25.

Note 1 Methods of heating for distillations are discussed in Experiment 3.

Note 2 Concentrated sulfuric acid (5 mL) may be substituted for the phosphoric acid. Using phosphoric acid is somewhat slower and safer; using sulfuric acid is faster but is more likely to lead to charring of the reaction mixture. If sulfuric acid is used, toward the end of the distillation be alert for the appearance of white fumes. If these appear, stop the heating at once.

Note 3 Hasten the drying process by swirling the flask occasionally. If there is insufficient time to complete the distillation in the same laboratory period as that in which the experiment was begun, leave the crude product over calcium chloride until the next laboratory period.

Note 4 The cotton plug should be no larger than the tapered end of a sharpened pencil.

Note 5 Methods of calculating yields are discussed beginning on page 8.

Preparation of Cyclohexene

Reaction equation

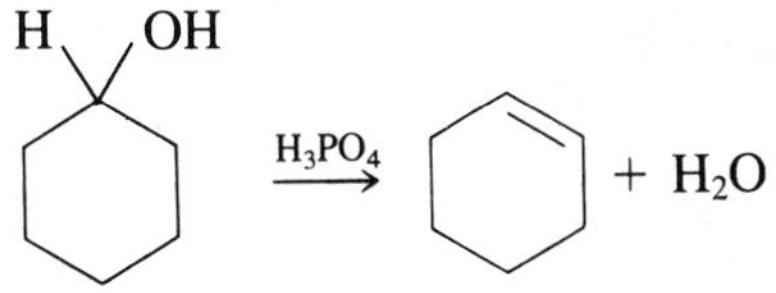

$$\text{Cyclohexanol} \xrightarrow{\text{H}_3\text{PO}_4} \text{Cyclohexene} + \text{H}_2\text{O}$$

Cyclohexanol
(Mol. Wt. 100)

Cyclohexene
(Mol. Wt. 82)

Amount of reactant used[1] = 20.0 g. (21 mL)

Moles of reactant = _____

Theoretical number of moles product obtainable = _____

Amount of product theoretically obtainable = _____

Theoretical yield _____ g

Actual yield _____ g

Percentage yield = _____ %

Results of tests with (a) Br_2—CH_2Cl_2 solution; (b) $KMnO_4$

(a) _____

(b) _____

---

[1]As cyclohexanol is the only reactant it is the limiting reactant.

1. What alcohol would be the most logical choice as starting material for the preparation of 1-methylcyclohexene? Write the equation for the reaction. Explain why you think your choice is the most logical.

2. How much of the alcohol you chose in answering Question 1 would you require as starting material if you needed to prepare 25 g of 1-methylcyclohexene and if your actual percentage yield were going to be 75%?

3. What is the theoretical yield of 1,2-dibromocyclohexane for the reaction of 20 g of cyclohexene with 20 g of bromine? (*Hint:* Which is the limiting reactant?)

4. A reaction is carried out for which the stoichiometry is $A + 2B + 3C \rightarrow D + 2E + F$. In this reaction 0.5 mole of A, 0.75 mole of B, and 1.0 mole of C are allowed to react. The yield of E is 0.2 mole. What is the percentage yield?

5. Consider the following reaction:

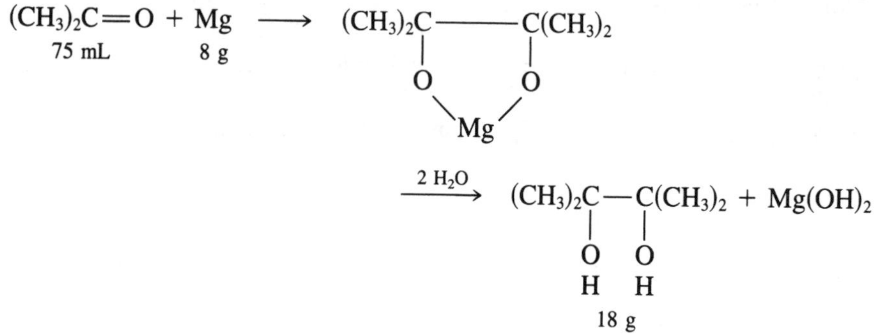

$$(CH_3)_2C=O + Mg \longrightarrow (CH_3)_2C-C(CH_3)_2$$

$$\underset{75 \text{ mL}}{} \quad \underset{8 \text{ g}}{}$$

with O—Mg—O bridge

$$\xrightarrow{2 \text{ H}_2O} (CH_3)_2C-C(CH_3)_2 + Mg(OH)_2$$

with O O / H H

18 g

Density of acetone = 0.7908

Which is the limiting reagent in this reaction? What is the percentage yield?

Preparation of Alkyl Halides

Alkyl halides are among the most important chemical intermediates used by the organic chemist. Although a number of alkyl halides have valuable practical applications as solvents, pesticides, and pharmaceuticals, many others are prepared principally for use as intermediates or starting materials in the synthesis of more complex molecules. For this reason, the organic chemist has developed a number of methods of synthesizing alkyl halides from readily available materials. This experiment illustrates the preparation of alkyl halides from alcohols, one of the most useful methods for making alkyl halides.

Replacement of the hydroxyl group of an alcohol by halogen may be carried out by treating the alcohol with a phosphorus halide, thionyl chloride, or with a halogen acid.

$$3 \text{ R—OH} + PX_3 \longrightarrow 3 \text{ R—X} + H_3PO_3 \quad (X = Br, Cl)$$

Phosphorus   Phosphorous
trihalide   acid

$$\text{R—OH} + PCl_5 \longrightarrow \text{R—Cl} + POCl_3 + HCl$$

Phosphorus   Phosphorus
pentachloride   oxychloride

$$\text{R—OH} + SOCl_2 \longrightarrow \text{R—Cl} + SO_2 + HCl$$

Thionyl
chloride

$$\text{R—OH} + HX \longrightarrow \text{R—X} + H_2O \quad (X = Br, Cl, I)$$

The choice of method to be used depends on many factors: cost, convenience, ease of purification, and, perhaps most important, the effect of the structure of the alcohol or the alkyl halide on the course of the reaction. In this experiment the replacement of the hydroxyl group is brought about through use of a halogen acid.

The replacement of a hydroxyl group by halogen using a halogen acid is an example of a nucleophilic substitution reaction in which one nucleophile is substituted for another. The most obvious exchange of nucleophilic groups, halide ion for hydroxide ion, is an experimentally impractical one because the equilibrium lies far to the left as written.

$$\text{R—OH} + X^- \;\rightleftharpoons\; \text{R—X} + OH^- \qquad \text{(Equilibrium far to left)}$$

Indeed, the reverse reaction is used to prepare alcohols from alkyl halides. In general, it may be said that hydroxide ion is a poor leaving group. In the presence of a strong acid, however, the alcohol is protonated to some extent on the oxygen atom to form an oxonium ion, and a much more useful exchange can take place, halide ion for the water molecule. Small, neutral molecules, such as the water molecule, are often excellent leaving groups.

$$R\!-\!OH + H^+ \rightleftharpoons R\!-\!\overset{+}{O}H_2$$
$$R\!-\!\overset{+}{O}H_2 + X^- \rightleftharpoons R\!-\!X + H_2O \quad \text{(Position of equilibrium dependent on } R\!-\!, X\!-\!, \text{ and conditions)}$$

In the acid-catalyzed reaction the position of the equilibrium is strongly dependent on the structure of the alkyl group ($R\!-\!$), the halogen ($X\!-\!$), and the conditions of the experiment, but generally conditions can be found in which the equilibrium is well toward the right.

The foregoing equations describe only the gross overall changes taking place during the reaction. The detailed description of the reaction pathway being followed, that is, the reaction mechanism, indicates the reaction to be of the S_N1 or S_N2 type. (In order to better understand this experiment, read the discussion of S_N1 and S_N2 reactions in your textbook.)

A. Preparation of *n*-Butyl Bromide

In the first part of this experiment *n*-butyl alcohol is treated with hydrobromic acid to produce *n*-butyl bromide according to the reaction equation below. The hydrobromic acid is prepared *in situ* from sodium bromide and sulfuric acid.

$$NaBr + H_2SO_4 \longrightarrow HBr + NaHSO_4$$

$$CH_3CH_2CH_2CH_2OH + HBr \longrightarrow CH_3CH_2CH_2CH_2Br + H_2O$$
n-Butyl alcohol   *n*-Butyl bromide

The conversion of *n*-butyl alcohol to *n*-butyl bromide is an excellent example of a reaction proceeding largely by an S_N2 mechanism.

$$CH_3CH_2CH_2CH_2\!-\!OH + H^+ \longrightarrow CH_3CH_2CH_2CH_2\!-\!\overset{\oplus}{O}H \ | \ H$$
n-Butyl alcohol

$$Br^- + CH_3CH_2CH_2CH_2\!-\!\overset{\oplus}{O}H \ | \ H \longrightarrow CH_3CH_2CH_2CH_2\!-\!Br + H_2O$$
n-Butyl bromide

However, when a primary alcohol is heated in the presence of a strong mineral acid, dehydration (Experiment 22) by an E1 or E2 mechanism also is possible. It is inevitable, therefore, that elimination occurs as a competing side reaction to produce some 1-butene.

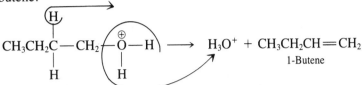

$$CH_3CH_2C-CH_2-\overset{\oplus}{O}-H \longrightarrow H_3O^+ + CH_3CH_2CH=CH_2$$

1-Butene

Fortunately, the primary alcohols are more resistant to dehydration than are the secondary and tertiary alcohols, and an olefin or an ether, if formed, is easily separated from the desired product by extraction with sulfuric acid.

Precautions.   *The following experiments require the use of hydrochloric and sulfuric acids and the handling of alkyl halides. See HAZARD CATEGORIES 2 and 4, page 4.*

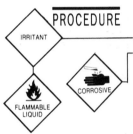

PROCEDURE   Place 31.0 g of sodium bromide, 35 mL of water, and 18.5 g (23 mL, 0.25 mole) of *n*-butyl alcohol in a 250-mL round-bottomed flask. Cool the mixture in an ice-water bath. Add 27 mL of concentrated sulfuric acid in 2- to 3-mL portions with thorough mixing (swirling) and cooling. Add a Boileezer, attach a reflux condenser (Figs. 0.1, 0.2, or 0.3; pages 25–29) and heat the mixture to boiling. Maintain a vigorous reflux for 45 minutes. Discontinue heating and cool the reaction mixture by immersing the flask in a cold-water bath. Remove the reflux condenser and fit the boiling flask with a connecting adapter and a condenser set for distillation (Fig. 3.1). (Note 1) Add a Boileezer and distill. The distillate will be composed of two layers—aqueous and organic. Continue distillation until the temperature reaches 110–115°. The upper oily layer in the distillation flask will have disappeared by this time and only water will distill over. You can verify this by collecting a few drops of distillate in a small (10 × 75 mm) test tube half-filled with water and noting whether the distillate contains droplets of oil or is composed only of water. Transfer the distillate to a separatory funnel and add 50 mL of water. Allow two layers to separate; withdraw and save the lower layer. Discard the upper water layer.

Caution!   *In the next step you will wash the product in a separatory funnel with sulfuric acid, then with water, then with aqueous sodium carbonate. Whenever a sodium bicarbonate or carbonate wash follows an acid wash, some effervescence is to be expected because of the liberation of carbon dioxide. Therefore, swirl the separatory funnel several times to obtain a preliminary mixing before you stopper and shake it. Quickly invert the funnel and release the pressure. Shake very briefly at first, releasing the pressure often.*

Return the crude *n*-butyl bromide to your separatory funnel and successively wash it with 25 mL of cold concentrated sulfuric acid, 25 mL of water, and 25 mL of 10%

— sodium carbonate solution. Allow ample time after each wash for a clean separation and be careful to save the proper layer. Separate the "wet" *n*-butyl bromide into a 50-mL Erlenmeyer flask and add approximately 1–2 g of anhydrous calcium chloride. Swirl the flask occasionally to hasten the drying process. When the product is clear, transfer it by way of a long-stemmed funnel fitted with a small cotton plug into a 50-mL distilling flask, add a Boileezer, and distill. Collect the material boiling in the range of 98–102°. The boiling point of pure *n*-butyl bromide is 101°. Yield, 20–25 g (58–74%). Record your yield on the report form and submit your product in a clean, labeled bottle to your laboratory instructor. Your product will be required in Experiment 24.

Note 1   If this is your first distillation, read Experiment 3 before carrying out the distillation.

B.   Preparation of *tert*-Butyl Chloride

Although hydrobromic acid reacts with primary, secondary, and tertiary alcohols to form alkyl bromides fairly readily, hydrochloric acid reacts satisfactorily only with the more reactive alcohols: tertiary, allylic, and benzylic. In terms of the general discussion given above, this observation can be rationalized by stating that chloride ion is not as good a nucleophile as bromide ion. The lesser nucleophilicity of the chloride ion can be compensated for, in part, by employing a better leaving group. Thus, in some preparations of alkyl chlorides, zinc chloride (a Lewis acid) is added to change the nature of and increase the effectiveness of the leaving group. This is the basis of the Lucas Test [Experiment 12 (C)] for distinguishing between primary, secondary, and tertiary alcohols.

   tert-Butyl alcohol reacts so readily with concentrated hydrochloric acid that you need not use zinc chloride in this experiment. The reason for this greater reactivity on the part of *tert*-butyl alcohol is explainable by the inductive and resonance stabilization of the intermediate carbocation that forms in the first stage of the reaction as indicated.

I
$$\underset{\overset{|}{CH_3}}{\overset{\overset{CH_3}{|}}{CH_3-C-OH}} + HCl \longrightarrow \underset{\overset{|}{CH_3}}{\overset{\overset{CH_3}{|}}{CH_3-C^{\oplus}}} + Cl^- + H_2O$$

$$\underset{\overset{|}{CH_3}}{\overset{\overset{CH_3}{|}}{CH_3-C^{\oplus}}} \longleftrightarrow \underset{\overset{|}{CH_3}}{\overset{\overset{H^{\oplus}CH_2}{\|}}{CH_3-C}} \longleftrightarrow \underset{\overset{|}{CH_3}}{\overset{}{H^{\oplus}CH_2=C}} \longleftrightarrow \underset{\overset{\|}{H^{\oplus}CH_2}}{\overset{\overset{CH_3}{|}}{CH_3-C}}$$

II
$$(CH_3)_3C^{\oplus} + Cl^- \longrightarrow \underset{\overset{|}{CH_3}}{\overset{\overset{CH_3}{|}}{CH_3-C-Cl}}$$

The positive carbocation unites in the second stage of the reaction with the negative chloride ion. The preparation of *tert*-butyl chloride is an excellent illustration of a reaction proceeding principally by the S_N1 reaction mechanism.

PROCEDURE Place 65 mL of concentrated hydrochloric acid (density 1.19, 36–38%) in a 125-mL Erlenmeyer flask and cool it to 0–5° in an ice bath. Transfer the cooled acid to a 125-mL separatory funnel. Add 15 g (19 mL, 0.2 mole) of *tert*-butyl alcohol to the funnel (Note 1) and shake the mixture occasionally for an interval of 15 minutes. Release the internal pressure regularly during the shaking period. Allow the mixture to stand in the funnel until the layers have separated. Using the technique described in Experiment 5, draw off the lower layer through the stopcock (after testing it to make certain that it is the water layer) and discard it.

Caution! *Before proceeding with the next step, read the cautionary note in the previous procedure regarding the possibility of effervescence and the formation of pressure in the separatory funnel when washing an acidic solution with sodium carbonate.*

Wash the crude product while it is still in the funnel by shaking it successively with 15 mL of water and with 15-mL portions of 5% sodium bicarbonate solution until the last traces of hydrochloric acid are neutralized. Wash once more with 15 mL of water. After each washing, withdraw the lower layer, test it, and discard it only if it is an aqueous layer. Transfer the crude *tert*-butyl chloride to a dry 50-mL Erlenmeyer flask and add 2 g of anhydrous calcium chloride. While the liquid is drying (Note 2), set up a distillation apparatus such as that shown in Figure 3.1, page 56, using a 50-mL distilling flask. The crude *tert*-butyl chloride, after drying over calcium chloride for the length of time required for you to assemble your distillation apparatus, may be filtered directly into your distillation flask. Use a long-stemmed funnel fitted with a *small* cotton plug to remove the particles of calcium chloride. Add 2–3 Boileezers or small pieces of clay plate and distill the product using an oil bath, an electric mantle, or a low flame. (Note 3) Collect the fraction boiling between 45–52° in either a weighed 50-mL Erlenmeyer or round-bottomed flask. (Note 4) Weigh your product and calculate your yield. Submit your sample to your instructor with your report.

Note 1   Tertiary butyl alcohol has a melting point of 25° and often is solidified in the bottle. It may be melted by placing the bottle in a warm water bath.

Note 2   Hasten the drying process by periodically shaking the alkyl halide and the drying agent.

Note 3   If this is your first distillation, read Experiment 3 before carrying out the distillation.

Note 4   The major portion of product will distill at the lower end of this temperature range.

C. Classification Tests for Alkyl Halides

In this part of the experiment the alkyl halides prepared in Procedures A and B are tested for their reactivity toward two classification reagents, silver nitrate in ethanol solution and sodium iodide in acetone solution. In general, ethanolic silver nitrate tends to react by the S_N1 mechanism with alkyl halides to form an alkyl nitrate and an insoluble silver halide.

$$R—X + AgNO_3 \longrightarrow R—ONO_2 + AgX$$

The order of reactivity of saturated, acyclic alkyl halides toward ethanolic silver nitrate is found to be:

Tertiary > Secondary > Primary

Tertiary halides generally react with immediate precipitation at room temperature. Primary and secondary halides react slowly, if at all, at room temperature but react readily at the boiling point of ethanol to give a precipitate. In general, sodium iodide in acetone tends to react with alkyl halides by the S_N2 mechanism to form the alkyl iodide and the insoluble (in acetone) sodium bromide or sodium chloride.

$$R—X + NaI \longrightarrow R—I + NaX$$

The order of reactivity of saturated, acyclic alkyl halides toward sodium iodide in acetone is found to be:

Primary > Secondary > Tertiary

With this reagent, primary bromides give a precipitate of sodium bromide within 3 minutes at room temperature, whereas primary chlorides must be heated to 50° to bring about a reaction. Secondary chlorides and secondary and tertiary bromides react at 50°, but tertiary chlorides react too slowly to give a positive test.

By employing both of these reagents, you can obtain valuable information about the structure of an alkyl halide.

PROCEDURE Place 2 mL of a 2% ethanolic silver nitrate solution (**Caution!** *Highly toxic.*) in each of two test tubes. Add one drop of your *n*-butyl bromide to one test tube and one drop of your *tert*-butyl chloride to the other test tube. Note and record whether or not a precipitate is formed in either tube. If no reaction is observed after 5 minutes, heat the solution to boiling in the steam bath. (**Caution!** *Ethanol is flammable.*)

Place 1 mL of the acetone solution of sodium iodide in each of two *dry* test tubes. Add one drop of your *n*-butyl bromide to one test tube and one drop of your *tert*-butyl chloride to the other test tube. Shake each tube to mix the contents, and allow the solutions to stand for 3 minutes. Note and record whether or not a precipitate is formed in either tube. If no precipitate has formed, place that tube in a beaker of water heated to 50°. (**Caution!** *Acetone is flammable; turn off burner before placing the test tube in the water bath.*) Allow the solution(s) to stand in the hot water for 6 minutes. Remove the test tube(s) and cool to room temperature. Note and record whether or not a precipitate is formed.

If the results obtained in either the silver nitrate or sodium iodide tests appear to be inconsistent with the general statements made in the introduction to this part of the experiment, repeat the tests with *n*-butyl bromide and *tert*-butyl chloride taken from the reagent shelf.

Name   *Section*   *Date*

Preparation of *n*-Butyl Bromide

Reaction equation $CH_3CH_2CH_2CH_2OH + HBr \longrightarrow CH_3CH_2CH_2CH_2Br + H_2O$

Quantities   18.5 g (23 mL)   ————

Mol. Wt.   ————   ————

Moles   ————   ————

Equivalents   ————   ————

Theoretical yield ———— g

Actual yield ———— g

Percentage yield ————

Preparation of *tert*-Butyl Chloride

Reaction equation $CH_3-\underset{\underset{CH_3}{|}}{\overset{\overset{CH_3}{|}}{C}}-OH + HCl \longrightarrow CH_3-\underset{\underset{CH_3}{|}}{\overset{\overset{CH_3}{|}}{C}}-Cl + H_2O$

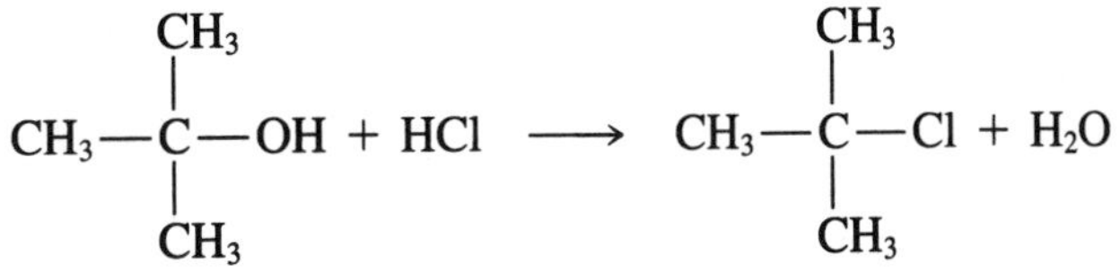

Quantities   15 g (19 mL)   ————

Mol. Wt.   ————   ————

Moles   ————   ————

Equivalents   ————   ————

Theoretical yield ———— g

Actual yield ———— g

Percentage yield ————

Classification Tests

Record the appearance of a precipitate with a plus (+) and the absence of a precipitate with a minus (−).

Compound	Silver nitrate test		Sodium iodide test	
	Unheated	Heated	Unheated	Heated
Your *n*-butyl bromide				
Reagent *n*-butyl bromide				
Your *tert*-butyl chloride				
Reagent *tert*-butyl chloride				

1. Write *balanced* equations to show how a sulfuric acid wash extracts the byproducts that inevitably result when *n*-butyl bromide is prepared by the method used in this experiment.

2. What products would you predict if isopropyl alcohol rather than *n*-butyl alcohol had been the starting material in this experiment?

3. Devise a procedure for the preparation of *n*-butyl bromide other than the one used in this experiment but one that begins with the same reagents.

4. What impurity could cause either *n*-butyl bromide or *tert*-butyl chloride, prepared as in this experiment, to give a false positive test with the two classification reagents?

5. Outline a procedure for the preparation of *sec*-butyl chloride. Include steps for the removal of impurities.

The Grignard Reaction: Preparation of 2-Methyl-2-Hexanol

The preparation of a Grignard reagent is a very rewarding experience for every organic chemistry student. It is a reaction that frequently tries one's patience, for Grignard reagents do not always form immediately, and often require some coaxing. It is especially important that glassware and all reagents be completely dry, because moisture will destroy the Grignard reagent according to the following equation:

$$RMgX + H_2O \longrightarrow RH + Mg(OH)X$$

It is advantageous to dry glassware in a drying oven overnight before use, if possible. All reagents must be anhydrous. Once the Grignard reagent is prepared it should be used without delay and with a minimum exposure to air.

PROCEDURE **_Caution!_** *Both diethyl ether and acetone are very flammable. Extinguish all flames in the vicinity while transferring these liquids and while using ether for extraction.*

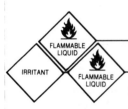

Arrange on *only one* ring-stand an assembly such as that illustrated in Figure 24.1 using a *dry*, 250-mL round-bottomed flask. (Note 1) Place in the flask 2.4 g (0.10 gram-atom) of magnesium turnings and 15 mL of anhydrous ether. Weigh in a clean, *dry*, 50-mL Erlenmeyer flask 13.7 g (10.75 mL, 0.10 mole) of previously prepared *n*-butyl bromide and transfer it to the separatory funnel. Fill the 50-mL Erlenmeyer flask with anhydrous ether and add it also to the dropping funnel. Swirl the funnel to ensure complete mixing of the *n*-butyl bromide and ether. Add in one portion approximately 15 mL of the ether solution of *n*-butyl bromide from the funnel to the magnesium turnings. Reaction usually begins within a few minutes and is accompanied by a spontaneous change in appearance from clear to an opalescent white and a gentle boiling of the ether. Add the remainder of the reagent from the separatory funnel dropwise to maintain a fairly rapid reflux. If the reaction becomes too lively, stop adding the reagent and immerse the round-bottomed flask in a cold-water bath. Too rapid reflux will cause ether vapor to escape from the top of the condenser. If the reaction does not begin spontaneously, warm the flask by immersing it in a warm-water (50°) bath. It may be necessary to gently crush the magnesium turnings with a large glass stirring rod slightly flattened on one end. This

---

Figure 24.1   Assembly for the preparation of a Grignard reagent. Notice that the system is open to the atmosphere.

Indicates placement of clamp

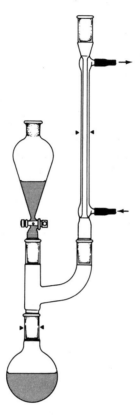

---

step ensures exposure of bare magnesium metal to the alkyl halide and usually initiates the reaction immediately. If this step is necessary you will need the help of your instructor or that of another student to lift the Claisen adapter with its attached dropping funnel and condenser clear of the reaction flask. Be sure to support the round-bottomed flask in your hand and make certain that there are no open flames nearby.

Once the reaction is in progress, the mixture becomes progressively darker. When refluxing subsides and nearly all of the magnesium has been consumed, immerse the reaction flask in a 50° water bath and allow the mixture to reflux for an additional 15–20 minutes. At the end of this time replace the warm-water bath with one containing cold water and cool the mixture. While the mixture is cooling prepare a solution of 5.8 g (7.4 mL, 0.10 mole) of dry acetone (Note 3) in 15 mL of anhydrous ether. Transfer the acetone-ether solution to the separatory funnel (be certain the stopcock is closed) and add it dropwise to the now-cold Grignard reagent, swirling the flask frequently by giving a rotatory motion to the ring-stand assembly. The reaction is very vigorous and good mixing is essential to ensure proper cooling. After adding all the acetone, allow the reaction mixture to stand at room temperature for 20–30 minutes. (Note 4) Hydrolyze the Grignard addition compound by stirring it very slowly into a cold mixture of 75 mL of saturated ammonium chloride and 25 g of crushed ice contained in a 250-mL beaker. Avoid transferring any unreacted magnesium. Rinse the reaction flask with a little ordinary ether and add the rinsings to the hydrolysis mixture. Stir until all solid material in the ether layer has dissolved. Transfer the mixture to a separatory funnel and separate. Extract the aqueous layer with a 25-mL portion of ordinary ether and add this ether extract to the separated ether solution. Discard the aqueous layer. Wash the ether once with 25 mL of 10%

IRRITANT

sodium carbonate solution and a second time with a 25-mL portion of saturated sodium chloride solution. Discard all aqueous washes. Dry the ether over anhydrous magnesium sulfate until the next laboratory period. Transfer the dried ether solution via a funnel fitted with a small cotton plug to a fractional distillation assembly (Fig. 4.1) and remove the ether using a hot-water bath. After all ether has been removed, the residue may be transferred to a smaller distillation assembly and the product distilled by heating with an electric mantle or over a very small flame. Collect as your desired product all material distilling in a temperature range of 137–143°. The boiling point of 2-methyl-2-hexanol is reported to be 143°. Yield, approximately 7.0 g (60%).

Note 1 The glassware must be dry. If it has not been dried in the oven overnight, it may be advisable to flame-dry the flask with a burner flame before assembling the apparatus, but do not light a burner if you or your neighbor are using ether.

Note 2 You can heat water for the hot-water baths in an ordinary pan or kettle on a hot plate, or by a burner in the fume hood or in a location well away from ether or acetone vapors. Since ether boils at about 35°, the water need not be heated above 50° except in the final step where the ether is being removed prior to the distillation of the product. If the bath is of a fairly generous size, it will not have to be refilled as often.

Note 3 Acetone should be dried over anhydrous magnesium sulfate at least overnight.

Note 4 You may stopper the reaction flask at this stage and work up the mixture the following laboratory period.

Name   Section   Date

The Grignard Reaction: Preparation of 2-Methyl-2-Hexanol

Reaction
equation

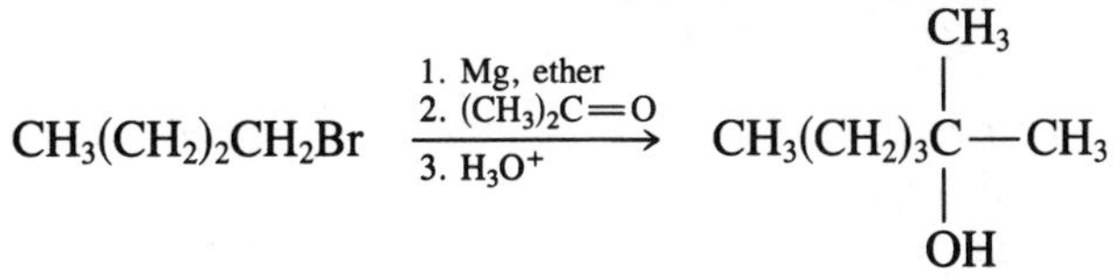

$$CH_3(CH_2)_2CH_2Br \xrightarrow[\text{3. } H_3O^+]{\substack{\text{1. Mg, ether} \\ \text{2. } (CH_3)_2C=O}} CH_3(CH_2)_3\overset{\overset{\displaystyle CH_3}{|}}{\underset{\underset{\displaystyle OH}{|}}{C}}—CH_3$$

n-Butyl bromide   2-Methyl-2-hexanol

Quantities   13.7 g (10.75 mL)   _____

Mol. Wt.   _____   _____

Moles   _____   _____

Equivalents   _____   _____

Theoretical yield _____ g

Actual yield _____ g

Percentage yield _____ g

QUESTIONS
and
EXERCISES

1. Some grades of ether may be anhydrous yet contain traces of ethyl alcohol. Would the presence of this impurity interfere in the preparation of a Grignard reagent? Explain.

2. Why is anhydrous magnesium sulfate rather than anhydrous potassium carbonate the preferred drying agent for acetone?

3. An attempted preparation of 2-methyl-2-hexanol appeared to proceed normally, but in the final distillation, after the ether had been removed, the only product recovered passed over at 85–90°. What product could this have been?

4. Write equations for two different preparations of 1-hexanol using a Grignard reagent.

5. Why are the Grignard preparations sometimes slow to start even when reagents are anhydrous and glassware dry?

6. The solvent of choice in a Grignard preparation is an anhydrous ether. Why could a Grignard not be prepared using an excess of the halogen compound as the solvent?

7. Using any readily available reagents and a Grignard reaction, show how the following alcohols might be prepared: (a) 3-methyl-1-butanol; (b) 2-phenyl-2-butanol; (c) 2,3,3-trimethyl-2-butanol.

8. The function of the ammonium chloride used to decompose the magnesium alkoxide formed in the addition of the Grignard reagent to the ketone is to provide a weakly acidic medium in which the magnesium salts will dissolve. Why is it better to use a weak acid such as ammonium chloride than a stronger acid such as hydrochloric acid?

9. When 1,2-dibromoethane is treated with magnesium, ethylene is formed rather than a Grignard reagent. The inorganic product is magnesium dibromide. Recalling that Grignard reagents resemble carbanions to some extent, write a mechanism for this elimination reaction.

Chromic Acid Oxidation: Preparation of Adipic Acid

Chromic acid, H_2CrO_4, is one of the most potent oxidizing agents the organic chemist uses. This reagent is generated by the treatment of either sodium dichromate or potassium dichromate with sulfuric acid according to the following equation:

$$Cr_2O_7^{2-} + 2\ H^+ \longrightarrow 2\ CrO_3 + H_2O$$

$$CrO_3 + H_2O \longrightarrow 2\ H^+ + CrO_4^{2-}$$

At room temperature chromic acid is capable of smoothly oxidizing a primary alcohol to a carboxylic acid and a secondary alcohol to a ketone and serves as a diagnostic test useful in the classification of alcohols (Exp. 10-D). Under more vigorous conditions chromic acid is capable of oxidizing an alkylated benzene to benzoic acid, effectively cleaving carbon-carbon bonds in the side chain down to the ring-attached carbon. The latter is oxidized to the carboxyl group, which is the highest oxidation state (3^+) permitted for one carbon atom attached to another.

Ring-opening oxidations leading to the formation of adipic acid may be accomplished using cyclohexanol or cyclohexanone as starting material. However, in both of these reactions the carbon atoms attacked are already bonded to oxygen. In the present experiment you will employ chromic acid to cleave the cyclohexene ring at its most vulnerable site—the carbon-carbon double bond.

Chromic acid oxidations are quite exothermic and careful control of the temperature must be maintained by the proper rate of addition of the oxidizing reagent. Should the reaction become too vigorous, fragmentation of the carbon chain and random oxidation may result.

Chromic acid is one of the most versatile oxidants available to organic chemists and is used in many forms and combinations. Unfortunately, hexavalent chromium compounds are thought to be carcinogenic. At present there are no satisfactory substitutes for the hexavalent chromium oxidizing agents; therefore, we must concentrate our efforts on learning to use them safely. Fortunately, few of them are volatile, so the principal hazard is contact with the body. Also, the trivalent chromium compounds are thought not to be carcinogenic; therefore, reduction of the hexavalent species is a relatively simple way to destroy and dispose of these materials.

> **Precautions.**   *This experiment requires the use of cyclohexene, concentrated sulfuric acid, and potassium dichromate. See HAZARD CATEGORIES 1, 2, and 4, page 4.*

PROCEDURE

Place 25 g of ice in a 500-mL Erlenmeyer flask and carefully add 30 mL of concentrated sulfuric acid. Mix by swirling the flask, then cool the acid solution in a cold-water bath. Clamp the flask in place. While the acid solution is cooling to room temperature, prepare in a 125-mL Erlenmeyer flask a second solution of 9 g of potassium dichromate ($K_2Cr_2O_7$) in 35 mL of water to which 10 mL of concentrated sulfuric acid has been added. Swirl until the dichromate has dissolved. Transfer the dichromate solution to a small separatory funnel and support the funnel on a ring-stand. Remove the flask containing the sulfuric acid solution from the cooling bath and add 4 mL (3.25 g, 0.04 mole) of cyclohexene. Swirl to dissolve as much of the hydrocarbon as possible. To the cyclohexene–sulfuric acid mixture add with swirling the potassium dichromate solution dropwise in approximately 1-mL increments. (Note 1) Keep the temperature of the reaction mixture at approximately 50°. The flask should feel quite warm but not too hot to hold. The addition should require 25–35 minutes. After all the dichromate solution has been added, the mixture should be a dark greenish blue. (Note 2) Support the flask on a steam bath or in a hot-water bath (85–90°) and warm for 30 minutes (**Hood!** *No flames!*). Periodically swirl the flask while heating. This step will help bring the reaction to completion and remove unreacted cyclohexene by evaporation. Next, pour the reaction mixture into a 400-mL beaker and allow the beaker to stand in an ice-water bath until the temperature is below 5°. Stir occasionally with a glass rod (not with your thermometer) and periodically scratch the walls of the beaker. When the temperature has reached 5°, you should see small crystals of adipic acid on the surface of the liquid. Collect the adipic acid in a small Büchner funnel, and wash with only a small volume (2–3 mL) of ice water. The crystals of adipic acid will be fairly pure as collected on the filter and will give a sharp melting point. Should your product have a slight greenish color, it may be recrystallized from a small volume of hot water.

Table 25.1   Solubility of Adipic Acid

Temperature, °C	g/100 g H_2O
15	1.44
40	5.12
50	9.24
60	17.6
70	34.1
100	100

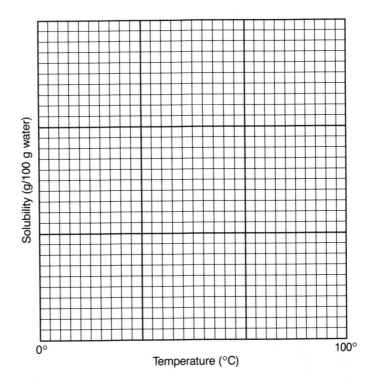

However, before recrystallizing your sample of adipic acid, use the values from the following solubility table to draw a solubility-temperature curve, and from this curve determine the volume of water you must employ to recover a maximum amount of product.

Cleanup: Your instructor will inform you of the proper method of disposal of the chromium-containing residues from your experiment.

Note 1 Reaction is immediate, as a color change and a temperature rise will indicate. If swirling is done above a white buret base or a piece of paper, you may note reaction progress by a sequential color change in the liquid from orange to green to blue.

Note 2 You may interrupt the procedure at this stage and allow the reaction mixture to stand overnight or until the following laboratory period. Actually, an improvement in the yield of adipic acid results from this additional reaction time.

Name _____   Section _____   Date _____

Preparation of Adipic Acid

$$3 \text{ (Cyclohexene)} + 4 \text{ K}_2\text{Cr}_2\text{O}_7 + 16 \text{ H}_2\text{SO}_4 \longrightarrow 3 \text{ HOOC}-(\text{CH}_2)_4-\text{COOH}$$

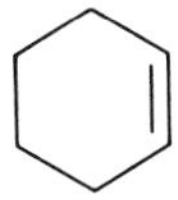

Cyclohexene

Adipic acid

$$+ 4 \text{ K}_2\text{SO}_4 + 4 \text{ Cr}_2(\text{SO}_4)_3$$
$$+ 16 \text{ H}_2\text{O}$$

	Cyclohexene	$K_2Cr_2O_7$	Adipic acid
Quantities	3.25 g	9.0 g	_____
Mol. Wt.	_____	_____	_____
Moles	_____	_____	_____
Equivalents	_____	_____	_____

Theoretical yield _____ g

Actual yield _____ g

Percentage yield _____

mp _____ °C

1. From the balanced reaction equation for the oxidation of cyclohexene given in the report form, determine the oxidation number change of the carbon atoms that were oxidized and the oxidation number change of the chromium atoms that were reduced.

2. What would the oxidation product have been if cyclohexene had been treated with ozone, followed by reductive hydrolysis?

3. A pseudoperspective drawing of the cyclohexene structure will show four of the ring carbon atoms in a tetrahedral bonding arrangement and four of the ring carbons lying in the same plane. Illustrate.

4. Write separate balanced oxidation-reduction reactions showing all products that result when the three isomeric butenes are treated with potassium dichromate and sulfuric acid.

5. If the solubility of adipic acid at $0°$ is 0.80 g per 100 g water, and the total yield of adipic acid obtained from the oxidation of a sample of cyclohexene amounted to only 0.87 g, what is the maximum volume of water that should be used to recrystallize the sample and recover 0.80 g of pure product?

6. Employing a different set of reagents for each reaction, show how the following compounds can be oxidized into the products shown:

   (a) toluene → benzoic acid

   (b) 2-butanone → propanoic acid

   (c) glucose → gluconic acid

   (d) 2,3-butanediol → acetaldehyde

Acid Derivatives:
Acyl Chlorides,
Anhydrides, and Amides

Usually acyl chlorides and acid anhydrides are prepared as intermediates in a reaction sequence, rather than as end products. As part of the identification or characterization of the acids, amides are often prepared as derivatives, although they are also important intermediates in a number of reaction sequences (see Experiment 36). Experiment 26 illustrates general methods commonly used for the preparation of all three classes of compounds. A general method for the preparation of an acyl chloride and an amide as a part of the characterization process is illustrated in Part I. The preparations of phenylacetyl chloride and phenylacetamide as intermediates in a reaction sequence are illustrated in Part II. Benzoic anhydride is prepared and purified in Part III.

Acyl chlorides are generally prepared by treating carboxylic acids with phosphorus trichloride, phosphorus pentachloride, or thionyl chloride.

$$3\ R-\overset{\displaystyle O}{\underset{\displaystyle OH}{C}} + PCl_3 \longrightarrow 3\ R-\overset{\displaystyle O}{\underset{\displaystyle Cl}{C}} + H_3PO_3$$

$$R-\overset{\displaystyle O}{\underset{\displaystyle OH}{C}} + PCl_5 \longrightarrow R-\overset{\displaystyle O}{\underset{\displaystyle Cl}{C}} + POCl_3 + HCl$$

$$R-\overset{\displaystyle O}{\underset{\displaystyle OH}{C}} + SOCl_2 \longrightarrow R-\overset{\displaystyle O}{\underset{\displaystyle Cl}{C}} + SO_2 + HCl$$

Each of these reagents has its advantages and disadvantages and no one of them can be used for all conversions of acids to acid chlorides. However, many organic chemists prefer the use of thionyl chloride, when it is applicable, because it is relatively simple to work with, is easily removed by distillation if employed in excess, and gives only volatile gases as byproducts. Thionyl chloride is less suitable than phosphorus trichloride for the preparation of low-boiling acyl halides because of difficulties in separating the product from the reagent and removing the dissolved gaseous byproducts.

The setup and procedure for Parts I and II are almost identical, except for reaction temperature and work-up. Usually the reaction is conducted at the boiling point of thionyl chloride (75–76°) for a short period of time; phenylacetyl chloride may give undesired byproducts at higher temperatures. Also, in many examples a small amount of dry cyclohexane is added at the end of the reaction before the evaporation of unreacted thionyl chloride and other volatile materials to help sweep out these substances; then the acyl chloride is purified by vacuum distillation (Experiment 29). For simplicity these useful additional steps have been omitted in this experiment.

You can prepare amides from acids (or acid derivatives) in a variety of ways (see discussion section of Experiment 30); however, one of the most common methods used is based on the reaction of an acyl chloride with ammonia or an amine.

$$R-C\begin{matrix}O\\\\Cl\end{matrix} + 2\ NH_3 \longrightarrow R-C\begin{matrix}O\\\\NH_2\end{matrix} + NH_4Cl$$

$$R-C\begin{matrix}O\\\\Cl\end{matrix} + 2\ NHR'_2 \longrightarrow R-C\begin{matrix}O\\\\NR'_2\end{matrix} + NH_2R'_2Cl$$

In general two equivalents of ammonia or amine are required, one being consumed in the neutralization of the hydrogen chloride liberated in the reaction. Reactions of acyl chlorides with amines are carried out by adding a solution of the acid chloride in methylene chloride to two equivalents of the amine in the same solvent. The insoluble amine hydrochloride generally precipitates from solution and may be removed by suction filtration. The amide is recovered by evaporation of the solvent. Reactions of acyl chlorides with ammonia can be carried out in the same fashion; however, this procedure is not convenient in the beginning organic laboratory. Therefore, in this experiment add the crude acyl chloride prepared in Procedure A to an excess of cold, concentrated ammonium hydroxide. The insoluble amide precipitates from solution. Preparations of amides in this fashion, although quite common, suffer from the disadvantage that part of the acyl chloride is hydrolyzed back to the acid, which dissolves in the basic solution. Therefore, yields by this procedure are often rather low, although the product is generally fairly pure.

PART I Preparation of *trans*-Cinnamamide

The following procedure is very general and is widely used for the preparation of amides from acids *via* the acyl chloride.

Cinnamic acid — $\xrightarrow{SOCl_2}$ — Cinnamoyl chloride — $\xrightarrow{NH_4OH}$ — Cinnamamide

| **Precautions.** *This experiment requires the use of liquid hydrocarbons, methanol, acyl halides, thionyl chloride, ammonium hydroxide, and pyridine. See HAZARD CATEGORIES 1, 2, 3, and 8, page 4.*

PROCEDURE   A.   Preparation of *trans*-Cinnamoyl Chloride

Caution! *Thionyl chloride is a low-boiling liquid (bp 75–76°) that is a strong irritant to the skin, eyes, and mucous membranes. It reacts vigorously with water to liberate hydrogen chloride and sulfur dioxide, which are also strong irritants. If the reagent comes into contact with the skin or clothing or is spilled on the bench, wash it off with copious quantities of water followed by dilute aqueous sodium bicarbonate. Thionyl chloride also attacks rubber; therefore, it is best handled in all-glass equipment with silicone grease on the ground joints. Handle it only in a good fume hood.* (Note 1)

Set up in the hood a reflux apparatus such as that shown in Figures 0.1, 0.2, or 0.3 (pages 25–29), using your smallest round-bottomed flask (25–100 mL). Temporarily remove the round-bottomed flask, weigh it, and place in it 3.0 g of *trans*-cinnamic acid, 2.5 g (1.5 mL) of pure thionyl chloride (Note 2), and two or three Boileezers. Immediately reassemble the apparatus. Heat the flask *gently* on the steam bath, in a heating mantle, or over a very low flame for 45 minutes. The evolution of gas should begin almost immediately, and the acid should dissolve in the thionyl chloride within the first 15 minutes of the heating period.

Allow the apparatus to cool. Replace the reflux condenser with a rubber stopper fitted with a short length of glass tubing bent to a right angle. Connect the tubing to a trap of the type shown in Figure 2.3 by means of a short length of suction hose. Connect the trap to the water aspirator and carefully reduce the pressure in the system to evaporate as much of the volatile material present as possible. When it appears that all of the easily removed volatile materials have been evaporated (about 5–10 minutes), release the vacuum, weigh the flask and crude product, and stopper the flask with the drying tube taken from the reflux condenser. Use the crude acid chloride immediately for the preparation of the amide.

Note 1   Use of a cork to attach the drying tube is acceptable. If the hood facilities are not adequate to serve the whole laboratory section, run the experiment on the open bench using a gas trap of either type (a) or (b) in Figure 35.1.

Note 2   Use the best available grade of thionyl chloride (e.g., Matheson Coleman and Bell reagent grade).

PROCEDURE B. Preparation of *trans*-Cinnamamide

Caution! *Because of the properties of the reactants, carry out the first part of this experiment (through the filtration step) in the exhaust hood.*

Place 30 mL of concentrated ammonium hydroxide in a 125-mL Erlenmeyer flask, stopper loosely with a clean rubber stopper, and chill the solution thoroughly in the ice bath. When the solution is ice cold, slowly add the *trans*-cinnamoyl chloride prepared in Procedure A of this experiment (Note 1), using a medicine dropper and stirring the mixture with each addition. There may be some sputtering and spattering as each drop reacts vigorously with the ammonium hydroxide. After the addition is complete, chill the mixture thoroughly (with occasional stirring) and collect the crude product on the Büchner funnel by suction filtration. Recrystallize the product from methanol (**Caution!** *Flammable!*) by dissolving it in the minimum amount of hot methanol and adding water as required. (Note 2) Collect the product by suction filtration, dry it thoroughly in the air on a clean piece of filter paper, weigh it, and determine its melting point. Yield, 1.4–2.4 g, mp 146–147°.

Note 1 If Procedure A was not assigned or if the *trans*-cinnamoyl chloride prepared appears to be of low quality, you may substitute 3.0 g of the commercially available acyl chloride.

Note 2 Because many amides are quite soluble in hot methanol, it may be necessary to add water dropwise to the hot solution just to the point where the solution becomes faintly turbid. If necessary, the hot solution may be clarified by adding a very small amount of methanol. Chill and collect the product. The use of excess pure solvent is sometimes necessary when the minimum volume of solvent is so small as to make further manipulations difficult. Adding a second poorer solvent then reduces the solubility to an acceptable level for a reasonable recovery of product. Avoid adding too large an initial excess of pure solvent, for it may then be difficult to get an acceptable recovery even by adding a second poorer solvent.

PART II   Preparation of Phenylacetamide

The following procedure, like that in Part I, is very general; however, because of the lower reaction temperature and longer reaction time it is usually employed where the higher temperature might lead to undesirable side products.

Phenylacetic acid $\xrightarrow{\text{SOCl}_2}$ Phenylacetyl chloride $\xrightarrow{\text{NH}_4\text{OH}}$

Phenylacetamide

PROCEDURE   A.   Preparation of Phenylacetyl Chloride

See **Caution** under Procedure A regarding the use of thionyl chloride.

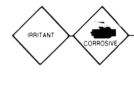

In the hood, set up a reflux apparatus such as that shown in Figures 0.1, 0.2, or 0.3 (pages 25–29), except that the top of the reflux condenser is fitted with a calcium chloride tube (Note 1), using your smallest round-bottomed flask (25–100 mL). (No source of heat is required for this experiment.) Temporarily remove the round-bottomed flask, weigh it, and place in it 2.8 g of phenylacetic acid (Note 2), 2.5 g (1.5 mL) of pure thionyl chloride (Note 3), and two or three Boileezers. (Note 4) Immediately reassemble the apparatus. The evolution of gas should begin almost immediately and the contents of the flask may become warm from the heat of reaction. Swirl the mixture carefully several times during the first 2 hours of reaction and then allow it to stand until the next laboratory period. (Note 5) Be sure to label your setup with your name to facilitate its identification.

Replace the reflux condenser with a rubber stopper fitted with a short length of glass tubing bent to a right angle. Connect the tubing to a trap of the type shown in Figure 2.3 by means of a short length of suction hose. Connect the trap to the water aspirator and cautiously reduce the pressure to evaporate as much of the volatile material present as possible. When it appears that all of the easily removed volatile materials have been evaporated (about 5–10 minutes), release the vacuum, weigh the flask and crude product, and stopper the flask with the drying tube taken from the reflux condenser. Use the crude acid chloride immediately in Procedure B.

Note 1   Use of a cork to attach the drying tube is acceptable. If the hood facilities are not adequate to serve the whole laboratory section, start the experiment on the open bench using a gas trap of either type shown in Figure 35.1. The setup can be transferred to the hood at the end of the period (3–4 hours) without water in the reflux condenser. In any of these setups use the drying tube during the time between laboratory periods.

Note 2   The lower aliphatic acids (butyric through capric) are liquids or low-melting solids that have distinctly unpleasant odors. Phenylacetic acid is also an aliphatic acid with an aromatic substituent (phenyl) on the α-carbon atom. It, too, has a very unpleasant odor. Therefore, take extreme care to keep these materials off of the skin, clothing, books, and laboratory bench. It is advisable to use plastic gloves of the throw-away type in working with these acids. Furthermore wash immediately after use or disassembly all glassware coming into contact with these acids. Very dilute (5%) aqueous sodium hydroxide followed by copious quantities of water or, for gummy or tarry residues, acetone followed by water will remove these materials very effectively from glassware or the laboratory bench. If any of these acids come in contact with the skin, wash them off immediately with soap and water. All of these substances are absorbed to some extent by the skin and cannot be completely washed off with soap and water. If you scrub well, any residual odor can be masked fairly well for several hours with a good spray deodorant. Usually the last traces of odor will be gone in 2–3 days.

The acid chlorides of these acids react with moisture to reform the acids; therefore, treat the acid chlorides with the same care.

Note 3   Use the best available grade of thionyl chloride (e.g., Matheson Coleman and Bell reagent grade).

Note 4   Boileezers are essential to prevent bumping.

Note 5   The water passing through the reflux condenser may be turned off at the end of the laboratory period (3–4 hours). If the standing period is less than several days, it may be necessary to complete the reaction by warming the flask with a warm water bath heated to 40–45° until gas evolution ceases. Do *not* overheat the mixture.

PROCEDURE   **B.   Preparation of Phenylacetamide**

Caution!   *Because of the properties of the reactants, carry out the first part of this experiment (through the filtration step) in the exhaust hood.*

Place 30 mL of concentrated ammonium hydroxide in a 125-mL Erlenmeyer flask and chill the solution thoroughly in the ice bath. When the solution is ice cold, slowly add the crude phenylacetyl chloride prepared in Procedure A of this experiment (Note 1), using a medicine dropper and stirring the mixture with each addition. There may be some sputtering and spattering as each drop reacts with the ammonium hydroxide. After the addition is complete, rinse the flask that contained the acid chloride with 5 mL of concentrated ammonium hydroxide and add to the Erlenmeyer flask. Chill the mixture thoroughly (with occasional stirring) and collect the crude product on the Büchner funnel by suction filtration. Recrystallize the product from methanol (*Caution!   Flammable!*) by dissolving it in the minimum amount of hot methanol and adding water as required. (Note 2) Collect the product by suction filtration, dry it thoroughly in the air on a clean piece of filter paper, weigh it, and determine its melting point. Yield, 1.4–1.8 g; mp 157–158°.

Note 1   If Procedure A was not assigned or if the phenylacetyl chloride prepared appears to be of low quality, substitute 3.0 g of the commercially available phenylacetyl chloride.

Note 2   Because the product is quite soluble in hot methanol, it will be necessary to add water dropwise to the hot solution just to the point where the solution becomes faintly turbid. If necessary, clarify the hot solution by adding a very small amount of methanol. Chill and collect the product. The use of excess pure solvent is sometimes necessary when the minimum volume of solvent is so small as to make further manipulations difficult. Adding a second poorer solvent then reduces the solubility to an acceptable level for a reasonable recovery of product. Avoid adding too large an excess of pure solvent, for it may then be difficult to get an acceptable recovery even by adding a second poorer solvent.

PART III   Preparation of Benzoic Anhydride

This experiment provides an illustration of a very general and effective method for the preparation of acid anhydrides from acid chlorides. Although this reaction is not often described in elementary texts on organic chemistry, it usually is the method of choice for the preparation of simple anhydrides.

Benzoyl chloride   **Pryidine**

Benzoic anhydride

Preparation of Benzoic Anhydride

Prepare a mixture of 2.0 mL (2.4 g, 0.017 mole) of benzoyl chloride and 8 mL of pyridine in a loosely stoppered 25-mL Erlenmeyer flask. Warm the mixture on the steam bath for 5 minutes and then pour it on 20 g of crushed ice and 10 mL of concentrated hydrochloric acid both in a 100-mL beaker. The anhydride separates at once; however, because of its low melting point it may separate as an oil. Cool the beaker containing the reaction mixture in ice and stir the mixture occasionally with a stirring rod or spatula. As soon as the product solidifies or crystallizes, collect it by suction filtration without allowing the solid to become warm, using a chilled funnel. (Note 1) Wash the product on the filter with two 50-mL portions of ice water and continue to pull air through the product for a few minutes to dry. Recrystallize the crude material from petroleum ether or hexane. (Note 2) Dry the product, weigh it, and determine its melting point. Yield, approximately 85–90%. Turn in the product with your report.

Note 1   If the material on the filter paper appears quite oily or mushy, washing with two 25-mL portions of petroleum ether may be helpful in the final crystallization.

Note 2   One challenging part of this experiment is the recrystallization of the product, mp 42–43°. Recrystallization of low-melting solids is more difficult than for higher melting compounds and requires patience, ingenuity, and skill. Often seeding the solution will aid crystallization, as will chilling in an ice-salt mixture or in Dry Ice-ethanol mixture. Add sufficient petroleum ether to form a homogeneous solution (one layer only) when the solution is warm.

Name   *Section*   *Date*

Preparation of *trans*-Cinnamamide

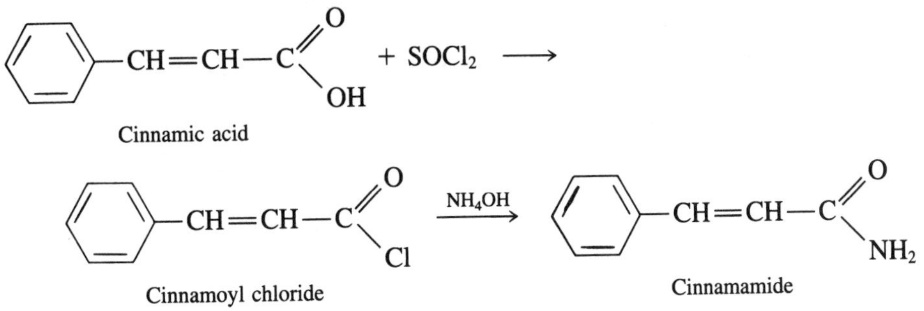

Cinnamic acid

Cinnamoyl chloride   Cinnamamide

	Acid	SOCl₂	Acyl chloride	Amide
Quantities	3.0 g	2.5 g	_____ g	_____ g
Mol. Wt.	_____	_____	_____	_____
Moles	_____	_____	_____	_____
Equivalents	_____	_____	_____	_____

Theoretical yield of acyl chloride _____ g

Actual crude yield of acyl chloride _____ g

Theoretical yield of amide _____ g

Actual yield of amide _____ g

Percentage yield of amide _____

Preparation of Phenylacetamide

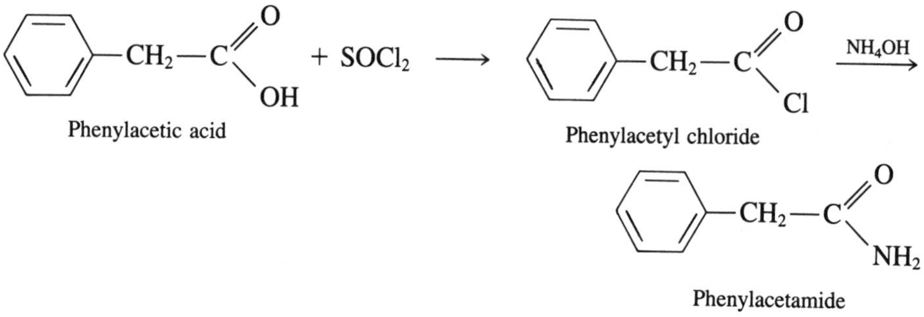

Phenylacetic acid   Phenylacetyl chloride

Phenylacetamide

	Acid	SOCl$_2$	Acyl chloride	Amide
Quantities	2.8 g	2.5 g	_____ g	_____ g
Mol. Wt.	_____	_____	_____	_____
Moles	_____	_____	_____	_____
Equivalents	_____	_____	_____	_____

Theoretical yield of acyl chloride _____ g

Actual yield of acyl chloride _____ g

Theoretical yield of amide _____ g

Actual yield of amide _____ g

Percentage yield of amide _____

Name ______ Section ______ Date ______

Preparation of Benzoic Anhydride

Reaction equation

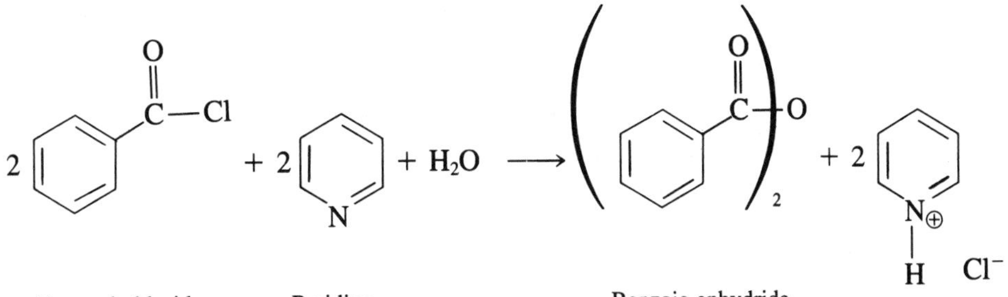

Benzoyl chloride | Pyridine | Benzoic anhydride

	Acyl chloride	Pyridine	Anhydride
Quantities	2.40 g	8 mL	______ g
Mol. Wt.	______	______	______
Moles	______	______	______
Equivalents	______	______	______

Theoretical yield ______ g

Actual yield ______ g

Percentage yield ______

1. Write equations for two different methods of synthesizing *trans*-cinnamamide that do not involve the reaction of an acyl chloride with ammonia.

2. What would be the product of the reaction of phenylacetyl chloride with (a) water, (b) *tert*-butylamine, (c) sodium acetate, (d) sodium methoxide, (e) bromine? Write the equations.

3. Write equations showing how phenylacetamide could be converted into (a) phenylacetonitrile, (b) benzylamine, (c) 2-phenylethylamine.

4. Acyl chlorides have higher molecular weights than the corresponding acids but are much more volatile (phenylacetic acid boils at 257° while phenylacetyl chloride boils at 210°). Why? Why is the difference in volatility less than that between corresponding alcohols and alkyl chlorides?

Esterification: Preparation of Isoamyl Acetate

Carboxylic acids react with alcohols to form **esters** according to the following equation:

$$R-\overset{\displaystyle O}{\overset{\displaystyle \|}{C}}-OH + R'-OH \rightleftharpoons R-\overset{\displaystyle O}{\overset{\displaystyle \|}{C}}-O-R' + H_2O$$

The esterification reaction involves an equilibrium that is attained only very slowly when the organic acid is allowed to react with the alcohol in the absence of a catalyst. The addition of a strong mineral acid such as sulfuric or hydrochloric acid, while catalyzing the reaction greatly, does not affect the position of the equilibrium. To force the reaction to the right (that is, to increase the amount of ester formed), two steps can be taken: (1) either reactant (usually the less costly) may be used in excess of the amount called for by the reaction equation, or (2) either of the products may be removed as it is produced. This experiment illustrates the preparation of a typical ester, isoamyl acetate. Isoamyl acetate (bp 142°), sometimes called banana oil, is a relatively common organic solvent used in paints and lacquers.

⟹ | **Precautions.** *This experiment requires the use of glacial acetic and sulfuric acids. See HAZARD CATEGORIES 2 and 8, page 4.*

PROCEDURE Preparation of Isoamyl Acetate

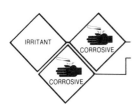

Set up a reflux apparatus such as that illustrated in Figures 0.1, 0.2, or 0.3 (pages 25–29), using a 125-mL round-bottomed flask. To the flask add 17.6 g (22 mL, 0.2 mole) of isoamyl alcohol ($d = 0.81$) and 30 g (30 mL, 0.5 mole) of glacial acetic acid ($d = 1.01$). Carefully add 4 mL of concentrated sulfuric acid to the mixture. Swirl the flask gently to thoroughly mix the reactants, add a Boileezer, and assemble the apparatus. Start the water flowing through the condenser and heat the mixture to boiling. If you use a burner, adjust it to a low flame. Allow the reaction mixture to reflux for one hour. Cool the solution in the reaction flask by immersing it in a cold-water bath, then pour it into 100 mL of cold water contained in a 400-mL beaker. Using a separatory funnel, separate the two layers. (Which is the ester layer?) Return the crude ester to the separatory funnel and wash it with 30 mL of 10% aqueous sodium bicarbonate solution, swirling the mixture in the funnel gently before shaking to permit any carbon dioxide formed to escape. (*Caution!* *Always make certain that any gas evolution expected is substantially complete before stoppering a separatory funnel. Release any pressure built up in the funnel at frequent intervals.*) Separate the layers and wash the ester layer by adding 30 mL of water and shaking the mixture for several minutes. Separate the layers and discard the aqueous layer. Transfer the ester to a 50-mL Erlenmeyer flask and dry it for 30 minutes over 3 g of anhydrous magnesium sulfate.

While the ester is drying, set up a distillation apparatus using a 100-mL distilling flask (Fig. 3.1). Make certain that all parts of the apparatus are dry (the condenser may still be wet from its use in the reflux apparatus) and see that all connections are tight. Filter the liquid into the distilling flask through a long-stemmed funnel fitted with a small plug of cotton. Add one or two Boileezers and distill the ester using a low flame, a heating mantle, or an oil bath. Collect that fraction boiling from 137–142° in a dry 50-mL flask. (Note 1) Weigh your product and submit it to your instructor with your report. Yield, approximately 75%.

Note 1 The vapor pressure of isoamyl acetate at room temperature is sufficiently high that, unless precautions are taken to keep exposure of the product to the air at a minimum, enough vapor will escape into the laboratory to produce an almost heady effect. Therefore, collect your product in a cooled receiver (Fig. 22.1) and lead vapors into the water trough.

Name ______ Section ______ Date ______

REPORT **27**

Preparation of Isoamyl Acetate

Reaction equation

$$\underset{\text{Acetic acid}}{CH_3\overset{\overset{O}{\|}}{C}OH} + \underset{\text{Isoamyl alcohol}}{(CH_3)_2CHCH_2CH_2OH} \rightleftharpoons \underset{\text{Isoamyl acetate}}{CH_3\overset{\overset{O}{\|}}{C}OCH_2CH_2CH(CH_3)_2} + H_2O$$

	Acid	Alcohol	Ester
Quantitites	30.0 g	17.6 g (22 mL)	______
Mol. Wt.	______	______	______
Moles	______	______	______
Equivalents	______	______	______

Theoretical yield ______ g

Actual yield ______ g

Percentage yield ______

1. Write a mechanism for the acid-catalyzed esterification reaction that will show the role of the sulfuric acid in the preparation of isoamyl acetate.

2. Although the addition of a small amount of sulfuric acid speeds up the attainment of equilibrium in acid-catalyzed esterification, the addition of a large amount of sulfuric acid can result in a slowing down of the attainment of equilibrium and a very low yield of ester. Suggest an explanation for this inhibitory effect of high acid concentration. (*Hint:* A clue may be found in the discussion preceding Experiment 23.)

3. An excellent acid catalyst for many esterifications is anhydrous hydrogen chloride, which is passed into a mixture of the reactants. Why would this catalyst be a poor choice for use with *tert*-butyl alcohol? What would be the expected product in such an attempted esterification?

4. The equilibrium constant for many acid-catalyzed esterifications is approximately 4, as is shown by the following expression.

$$K_e = \frac{[\text{products}]}{[\text{reactants}]} = \frac{[\text{ester}][\text{water}]}{[\text{acid}][\text{alcohol}]} \cong 4$$

Calculate the theoretical yield of isoamyl acetate using the amounts of isoamyl alcohol and acetic acid specified in the experiment and using 4 for K_e.

5. Propose a synthesis for isoamyl acetate that would be a one-way synthesis not involving any equilibrium between reactants and products.

Esterification:
Preparation of Aspirin

The phenols, unlike the alcohols, cannot be esterified by direct interaction of the phenol with an organic acid. The esterification of a phenol usually is carried out by treating it with an acid anhydride or an acyl chloride. In this experiment the phenolic acid, salicylic acid, is converted into the acetate ester, aspirin, by treatment with acetic anhydride.

| Salicylic acid | Acetic anhydride | Aspirin |

Precautions. *This experiment requires the use of acetic anhydride, phosphoric acid, and methanol. See HAZARD CATEGORIES 2, 3, and 8, page 4.*

PROCEDURE

If you are assigned both this experiment and the next (preparation of methyl salicylate), you can do them simultaneously by starting the methyl salicylate preparation first and then carrying out the aspirin preparation while the methyl salicylate reaction mixture is being heated under reflux for 3–3½ hours.

A. Preparation of Salicylic Acid Acetate

Place 2 g of salicylic acid in a 125-mL Erlenmeyer flask, add 5 mL of acetic anhydride and 5 drops of 85% phosphoric acid. Stir the mixture well. Heat the flask in the boiling-water bath for 5 minutes, remove from the bath and, while still hot, carefully add 2 mL of water in one portion. (**Caution!** *The solution may boil from the heat of decomposition of the excess acetic anhydride; handle the flask carefully.*) After decomposition is complete, add 40 mL of water and stir the solution until crystals begin to form. Cool the mixture in the ice bath to complete the crystallization. Collect the product by suction filtration on the Hirsch funnel, wash with 5 mL of cold water, and pull air through the filter until the product is dry. Recrystallize the product from 35 mL of hot water (Note 1), using decolorizing charcoal if the product is colored. Dry the product and determine its melting point. Perform the test described below and turn in your product together with your report.

B.   Test for the Phenolic Hydroxyl Group

Dissolve a few crystals of aspirin in 1 mL of methanol in a test tube and add one drop of 1% ferric chloride solution. Record your result (see page 151). Repeat the test with salicylic acid. Result?

Note 1   Do not heat the water above 80°. Boiling water will partially hydrolyze aspirin to salicylic acid and acetic acid.

Name Section Date

REPORT
28

Preparation of Aspirin (Salicylic Acid Acetate)

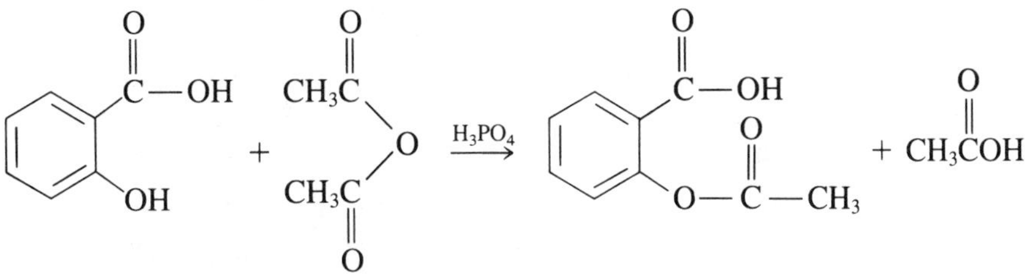

Salicylic acid   Acetic anhydride   Aspirin

	Salicylic acid	Acetic anhydride	Aspirin
Quantities	2.0 g	(5 mL) 5.41 g	________ g
Mol. Wt.	________	________	________
Moles	________	________	________
Equivalents	________	________	________

Theoretical yield ________ g

Actual yield ________ g

Percentage yield ________

Result of the ferric chloride test on aspirin ________________________

Result of the ferric chloride test on salicylic acid ________________________

1. Another pharmaceutical product derived from salicylic acid is salol (phenyl salicylate), the phenyl ester of salicylic acid. Since phenol cannot be esterified by direct interaction with salicylic acid, some indirect method must be used. Write equations for a sequence that might be used to prepare salol from phenol and salicylic acid.

2. Aspirin and salol are both acidic substances. Which is the stronger acid? Why do both substances pass through the stomach unchanged and first become hydrolyzed in the intestines?

3. Another anti-inflammatory pharmaceutical product is ibuprofen (Motrin), 2-(*p*-isobutylphenyl)propionic acid. Draw its structure. Suggest a possible synthesis of ibuprofen starting with ethylbenzene.

Esterification:
Preparation of
Oil of Wintergreen;
Vacuum Distillation

In this experiment salicylic acid is esterified with methanol at the carboxyl group to yield methyl salicylate (oil of wintergreen). Oil of wintergreen is an important pharmaceutical chemical.

$$\text{Salicylic acid} \quad + \quad CH_3\text{—OH} \quad \xrightarrow{H_2SO_4} \quad \text{Methyl salicylate}$$

This experiment also illustrates the use of distillation under reduced pressure for the purification of high-boiling liquids. Distillation of high-boiling liquids at atmospheric pressure is often unsatisfactory because at the high temperatures required the material being distilled may partially (or even totally) decompose, resulting in a loss of product and contamination of the distillate. Furthermore, unless the distillation apparatus is properly insulated, thermal losses to the room may be rather large, requiring that the liquid in the still pot be heated even more strongly and slowing down the rate of distillation. However, as we pointed out in the introduction to Experiment 3, when the total pressure (inside the distillation apparatus) is reduced, the boiling point is lowered. Therefore, by varying the pressure under which a distillation is conducted, the temperature required for distillation can be controlled. Distillations carried out in this fashion are called **vacuum distillations** or **distillations under reduced pressure.**

A simple apparatus setup for distillation under reduced pressure is shown in Figure 29.1. This setup requires the use of a Claisen adapter. In the first neck of the adapter is inserted a capillary ebullator, the function of which is to admit a *tiny* stream of air (or preferably nitrogen or helium) bubbles to prevent bumping during distillation. Boileezers do not function satisfactorily at reduced pressure. The rest of the setup is fairly typical of simple, atmospheric distillations except that the receiver is fitted with an adapter connected to some pressure-reducing device. Two means of reducing pressure are commonly employed in the organic chemistry laboratory, the water aspirator and the oil vacuum pump. The aspirator is useful down to pressures

Figure 29.1 Vacuum distillation assembly. A short section of rubber tubing should be stretched over the top of the capillary ebullator. A short length of copper wire inserted into the tubing will prevent a complete cut off of air when clamp is closed.

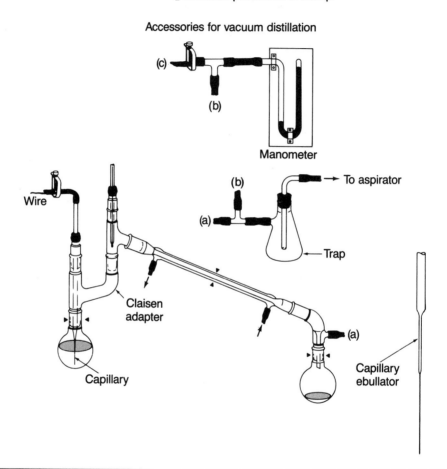

of about 12 Torr and the vacuum pump down to pressures below 1 Torr. Although the use of an oil pump is no more difficult than the use of an aspirator, it requires more care and rather costly installation; therefore, even in advanced laboratories the aspirator is widely used when the pressure requirement permits.

Some means of measuring the pressure inside the distillation apparatus must be provided. There are many devices available for such measurements, but either a simple manometer or the special manometer shown in Figure 29.1 is suitable for distillations at aspirator pressures. If needed, the apparatus can be modified so that the pressure can be controlled—by "bleeding" in a restricted amount of air at point (c) in the apparatus, for example—but in simple setups this feature is usually omitted.

At 760 Torr pressure methyl salicylate boils at 222.3°. With care it could be distilled at this pressure, but the process would be more demanding than a distillation under reduced pressure. Figure 29.2 shows the variation in the boiling point of methyl salicylate over the pressure range of about 7 Torr to 50 Torr. In the range of the water aspirator (12–20 Torr) methyl salicylate should distill in the very convenient temperature range of about 100–115°.

Carry out all experiments in the organic chemistry laboratory with due regard for the hazards involved. When a distillation is conducted under reduced pressure, treat the apparatus with more care than in a distillation at atmospheric pressure because of the possibility of an implosion if some part of the glassware in the system being evacuated is broken. Inspect glassware used in such systems for flaws or cracks that might rupture under the approximately 15 pounds per square inch pressure applied to the outer surface and essentially zero pressure on the inner surface. Most laboratory glassware designed for the organic chemistry laboratory and having cylindrical or spherical shapes (or nearly so) can be used safely. Many Erlenmeyer flasks are not designed for use under reduced pressure, and any such flask of over 50-mL capacity should not be evacuated unless it is known that the flask was intended for use under reduced pressure. If adjustments are required in a system being evacuated, allow the pressure to return to its normal value before moving, twisting, or otherwise applying force to the system (except for adjustable components built into the system such as stopcocks, fraction collectors, etc.). *Wear safety glasses while carrying out a vacuum distillation.*

Figure 29.2   The boiling point of methyl salicylate at various pressures.

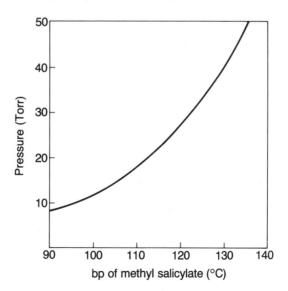

>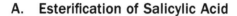
>
> **Precautions.**   *This experiment requires the use of salicylic acid, ether, methanol, and sulfuric acid. See HAZARD CATEGORIES 1, 2, and 3, page 4.*

PROCEDURE

A.   Esterification of Salicylic Acid

Set up a reflux apparatus such as that illustrated in Figures 0.1, 0.2, or 0.3 (page 25–29), using a 100-mL round-bottomed flask. To the flask add 6.9 g (0.05 mole) of salicylic acid and 24 g (30 ml, 0.75 mole) of methyl alcohol. *Carefully* add, in small portions, 8 mL of concentrated sulfuric acid. Swirl the flask gently after each addition of acid to thoroughly mix the reactants. Add a Boileezer to the flask, and assemble the apparatus. Start the water flowing through the condenser and heat the mixture to boiling, using a low flame, a heating mantle, or an oil bath. Allow the mixture to reflux for 3–3 $\frac{1}{2}$ hours. (Note 1) Cool the solution in the reaction flask by immersing it in a cold-water bath, then add 50 mL of water. Pour the mixture into a 125-mL separatory funnel. Rinse the reaction flask with 25 mL of ether and add the ether rinsings to the separatory funnel. Shake well and separate the layers. Extract the aqueous layer with a second 25-mL portion of ether and combine the ether layers. Discard the aqueous layer. Wash the ether solution once with 25 mL of a 5% $NaHCO_3$ solution and once with 25 mL of a saturated sodium chloride solution. Separate and store the ether solution in a 125-mL Erlenmeyer flask over 1–2 g of anhydrous calcium chloride until the next laboratory period.

B.   Distillation Under Reduced Pressure[1]

Caution!   *See last paragraph of discussion on distillation under reduced pressure, page 305.*

Set up an apparatus for distillation under reduced pressure by assembling the equipment illustrated in Figure 29.1, using a 50-mL round-bottomed flask fitted with a Claisen adapter. Fit the distilling flask with a capillary ebullator tube reaching to the bottom. You can make a capillary ebullator tube by heating a length of 6-mm glass tubing in your burner until it is very soft, removing it from the flame, and rapidly drawing it out to a diameter slightly smaller than that of a capillary melting point tube and an added length of 8–12 inches. Heat the capillary at a point about 2 inches from its end in a low flame and rapidly draw it out to a diameter about that of a hair. Break the fine capillary to a length of 2 inches or longer. (Note 2) If a

---

[1] The water pressure in some laboratories is greatly diminished when all students attempt to use aspirators at the same time. If vacuum pumps are available, some instructors may prefer to modify the procedure under Part B to adapt it to the use of a high vacuum device. In this event, the trap should be one cooled by Dry Ice and the distillation carried out using the combined ester residues of 5–6 students.

bleeder tube is supplied with the ground-glass equipment kit, it will serve as a capillary if equipped as illustrated in Figure 29.1. Connect the side-arm of the take-off adapter to the glass "tee" [labeled (a) in Fig. 29.1] leading to the manometer and trap and connect the trap to the water aspirator.

Introduce the dried ether solution of methyl salicylate into the distilling flask via a long-stemmed funnel. Make certain that all connections and screw clamps are tight. Place the apparatus under diminished pressure by turning on the water aspirator full force and remove the ether, using a warm-water (40°) bath as the heat source. (**Caution!** *No open flames.*) A thin stream of bubbles should come out of the ebullator tube. When you have removed all the ether (no more distillate appears to pass over), stop heating, open the screw clamp or stopcock at the manometer "tee" [labeled (c) in Figure 29.1], and turn off the water. Always break the vacuum before turning off the aspirator. Change receivers and resume the distillation under diminished pressure, distilling the ester using a low flame, heating mantle, or oil bath. Collect all distillate that passes over at a temperature of 100° or above. (Note 3) When no more distillate passes over, stop heating and open the apparatus to the atmosphere as described previously. Weigh the product in a dry 25-mL Erlenmeyer flask, label, and submit it to your instructor with your report. Yield of pure product, approximately 5.0 grams (66%).

Note 1   A long reflux time is required to esterify salicylic acid and obtain a respectable yield. A supplementary experiment may be performed during this reaction period, or work from previous experiments still pending may be completed. If necessary you may spread the reflux time over two laboratory periods.

Note 2   The capillary ebullator may be tested by blowing through it into a small quantity of acetone in a beaker. If the capillary is of the proper size you will see a thin stream of fine bubbles. If no bubbles form, the capillary is too small. If you see a coarse stream of bubbles, the capillary is too large.

Note 3   The normal boiling point of methyl salicylate is 224° but the ester will distill at 105° at a pressure of 14 Torr. A pressure of 14 Torr is about the limit obtainable with a good aspirator and a water temperature of 16.5°. If a different pressure is obtained, the corresponding temperature at which the ester will boil may be determined from the curve, Figure 29.2.

Name   Section   Date

Preparation of Methyl Salicylate (Oil of Wintergreen)

	Salicylic acid	Methanol	Methyl salicylate
Quantities	6.9 g	24 g	_____ g
Mol. Wt.	_____	_____	_____
Moles	_____	_____	_____
Equivalents	_____	_____	_____

Partners in experiment:

Theoretical yield _____ g   1. _____

Actual yield _____ g   2. _____

Percentage yield _____   3. _____

1. Many of the more expensive vacuum distillation setups avoid the use of a capillary ebullator, substituting a magnetic stirrer or a mechanical stirrer (driven by a magnet). What is one disadvantage of the capillary ebullator that could make it difficult to obtain a pure product by the vacuum distillation of certain compounds? How can we modify the ebullator shown in Figure 29.1 to avoid this difficulty?

2. For reasons of expedience, the distillation of methyl salicylate under reduced pressure may be carried out using a low flame as a heat source. Actually, you achieve a more accurate relationship between pressure and boiling point when heating with an electric mantle or an oil bath. Why?

3. What would be the pressure if methyl salicylate were brought to a boil at 140°?

The Preparation of Acetanilide

Ammonia reacts with acyl halides, anhydrides, or esters to yield the corresponding **amides.**

$$R—\overset{\overset{\textstyle O}{\|}}{C}—Cl + NH_3 \longrightarrow R—\overset{\overset{\textstyle O}{\|}}{C}—NH_2 + HCl$$

$$\begin{matrix} R—\overset{\overset{\textstyle O}{\|}}{C} \\ \diagdown \\ O + NH_3 \\ \diagup \\ R—\underset{\underset{\textstyle O}{\|}}{C} \end{matrix} \longrightarrow R—\overset{\overset{\textstyle O}{\|}}{C}—NH_2 + R—\overset{\overset{\textstyle O}{\|}}{C}—OH$$

$$R—\overset{\overset{\textstyle O}{\|}}{C}—OR' + NH_3 \longrightarrow R—\overset{\overset{\textstyle O}{\|}}{C}—NH_2 + R'—OH$$

If one or more of the hydrogen atoms of the ammonia molecule is replaced by an alkyl group, the resultant compound is an **amine.** Amines that have at least one hydrogen atom on the nitrogen atom, like ammonia, react with the above acid derivatives to yield **N-substituted amides.** N-substituted amides in which the hydrocarbon group on the nitrogen atom is a phenyl group, C_6H_5—, are called **anilides.** In the present experiment aniline is treated with acetic anhydride to produce acetanilide.

Aniline $+$ acetic anhydride $\longrightarrow$ Acetanilide $+ CH_3—\overset{\overset{\textstyle O}{\|}}{C}—OH$

Precautions. *This experiment requires the use of aniline, hydrochloric acid, and acetic anhydride. See HAZARD CATEGORIES 2, 3, and 8, page 4.*

PROCEDURE

Dissolve 20 mL of concentrated hydrochloric acid in 250 mL of water and add 20 g (20 mL, 0.222 mole) of aniline, stirring until the aniline dissolves (you may use a beaker for this step). If the aniline is badly discolored, add 2 g of decolorizing charcoal, stir the mixture for a few minutes, and filter with suction through the Büchner funnel. Dissolve 32 g of sodium acetate crystals (CH$_3$CO$_2$Na · 3 H$_2$O) in 100 mL of water. Filter the sodium acetate solution if necessary to remove undissolved particles.

Pour the solution of aniline hydrochloride into a 600-mL beaker and gently warm the solution on a wire gauze over a low flame or on the hot plate while stirring it with a thermometer. When the temperature of the solution reaches 50°, remove the beaker, place it on the desk, and add 24 mL of acetic anhydride. (**Caution!** *The vapors of acetic anhydride are very irritating to the nose, throat, and eyes.*) Stir the solution until the acetic anhydride dissolves and *immediately* add in one portion the solution of sodium acetate. Stir the mixture vigorously. Crystals of pure acetanilide should begin to precipitate almost immediately. Cool the reaction mixture in an ice bath and continue to stir the mixture while the product crystallizes. Collect the

crystals by filtering them with suction on the Büchner funnel, wash them with 20 mL of cold water, and press the crystal cake on the filter with a clean rubber stopper. Allow the crystals to dry until the next laboratory period. The acetanilide you have prepared may be used in following experiments without further purification. Protect the crystals from dust and dirt with a piece of clean filter paper or paper towel.

Weigh the dried product and determine its melting point. Transfer your product into a clean, dry, labeled bottle and turn it in with your written report. Your instructor will return your preparation for use in Experiment 36. Yield, 14–16 g; mp 115°.

Preparation of Acetanilide

Reaction
equation

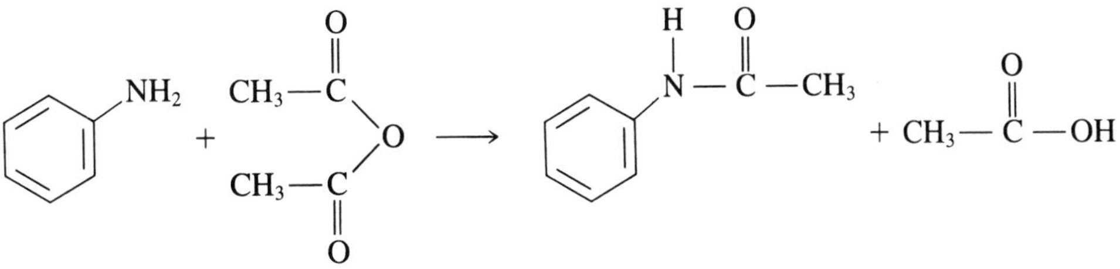

| | Aniline | Acetic anhydride | Acetanilide |

| Quantities | 20 mL (20 g) | 24 mL (26 g) | _____ g |

Mol. Wt. _____   _____   _____

Moles _____   _____   _____

Equivalents _____   _____   _____

Theoretical yield _____ g

Actual yield _____ g

Percentage yield _____

mp _____ °C

1. Although ethyl phenylacetate reacts readily with ammonia and with methyl-amine to yield the corresponding amide, the reaction of ethyl phenylacetate with *t*-butylamine is so slow and the yield so small as to render the reaction useless in the laboratory. Suggest a reason for the failure of the aminolysis with *tert*-butylamine. (*Hint:* What is the mechanism of aminolysis?)

2. Write equations for the syntheses of the following amides: (a) benzanilide, (b) N-*tert*-butylphenylacetamide, (c) succinimide.

3. What is the function of the sodium acetate used in the preparation of acetanilide according to the procedure given? Write the equation for the preparation of acetanilide as you have actually performed the experiment, *not the equation on the report form!* Include all reactants and indicate the role of sodium acetate.

4. Acetyl chloride and ketene, $H_2C{=}C{=}O$, may be used to acetylate aniline but both reagents offer certain disadvantages. Suggest possible disadvantages to the use of each. Do these reagents offer any obvious advantages?

5. For persons sensitive to aspirin, 4-hydroxyacetanilide (acetaminophen, Tylenol) often is prescribed as an analgesic. Suggest a synthesis leading to the preparation of this pharmaceutical product beginning with phenol or chlorobenzene.

The Friedel-Crafts Reaction: Preparation of *p*-Anisyl Benzyl Ketone or *p*-Methoxybenzophenone

Although the Friedel-Crafts reaction is a very general reaction in which an unsaturated hydrocarbon (alkene, alkyne, or aromatic hydrocarbon), or certain simple derivatives thereof, is alkylated or acylated using a Lewis acid catalyst, the most commonly encountered examples of this method involve the alkylation or acylation of aromatic hydrocarbons, aryl halides, or aryl ethers. A variety of alkylation agents may be used, alkyl halides, alcohols, and alkenes being the most common. Acyl halides and acid anhydrides are the most frequently employed acylation agents, but, as this experiment illustrates, under the proper conditions carboxylic acids may also be used. Aluminum chloride is probably the most common Friedel-Crafts catalyst and, for that reason, appears in most general equations for the Friedel-Crafts reaction. Other catalysts include zinc chloride, stannic chloride, sulfuric acid, and hydrofluoric acid. In practice the choices of aromatic substrate, alkylating or acylating agent, and catalyst are all interdependent, and the selection of one component in the reaction may limit possible choices for the remaining components.

This experiment illustrates the acylation of an aromatic ether (anisole) using a carboxylic acid (phenylacetic acid or benzoic acid) as the acylating agent and polyphosphoric acid (PPA) as the catalyst.

Phenylacetic acid   Anisole

The use of polyphosphoric acid as a Friedel-Crafts catalyst is a relatively recent innovation. In many reactions it has the advantage of serving as an effective acid catalyst, serving as solvent for the reaction, resulting in high yields with few byproducts, and simplifying the equipment setup and the work-up of product. In the more conventional Friedel-Crafts acylation shown in the following equation either an excess of the aromatic substrate or some inert organic compound must be used as solvent. The reaction is often heterogeneous, and you must make provisions for the handling of anhydrous aluminum chloride and the trapping of escaping hydrogen halides (from the aluminum chloride and/or the acylating agent). There is no real difficulty in carrying out reactions of this type; however, the simplicity and ease of carrying out the polyphosphoric acid-catalyzed Friedel-Crafts reaction recommends its use whenever it is applicable.

$$R-\overset{\overset{\textstyle O}{\|}}{C}-X + \langle\!\!\langle\ \rangle\!\!\rangle-R' \xrightarrow{\text{AlCl}_3} R-\overset{\overset{\textstyle O}{\|}}{C}-\langle\!\!\langle\ \rangle\!\!\rangle-R' + HX$$

$$X = Cl \quad \text{or} \quad R-\overset{\overset{\textstyle O}{\|}}{C}-O$$

Note that, whatever the catalyst, the predominant product in acylations of this type is the *para*-isomer. Generally, very little or none of the *ortho*- or *meta*- isomers is obtained.

Your instructor will assign to you either procedure A or B of this experiment. The method used in Procedure A is simpler, but it requires that you work with phenylacetic acid (Note 1), an important and relatively safe carboxylic acid but, unfortunately, one that shares with its low molecular weight aliphatic relatives (valeric through capric acid) a very unpleasant odor. The ability to work with phenylacetic acid without carrying its characteristic odor away from the laboratory is indicative of superior laboratory technique.

> **Precautions.** *This experiment requires the use of ether, methanol, and polyphosphoric acid. See HAZARD CATEGORIES 1, 2, and 3, page 4.*

PROCEDURE **Caution!** *Polyphosphoric acid is a very viscous liquid that is not easy to pour, particularly when cold. When in use, keep it in a reasonably warm place and handle it with care. If any of the acid is spilled on the bench or comes into contact with any part of the body or clothing, remove it at once by washing the affected area with a large volume of water, followed by dilute aqueous sodium bicarbonate, and more water. Phenylacetic acid has a persistent, unpleasant odor; therefore, take care to avoid getting it on the skin or clothing. (Note 1)*

A.   Preparation of *p*-Anisyl Benzyl Ketone

In a 125-mL flask place 15 g of polyphosphoric acid, 1.4 g of phenylacetic acid (Note 1), and 1.2 g of anisole. Loosely stopper the flask with a clean rubber stopper and place it in a boiling-water bath. (Note 2) After a minute or two remove the flask and swirl it vigorously to mix the contents and obtain a homogeneous reaction mixture. Return the flask to the bath and allow it to remain there for 45 minutes, except for brief periods at about 10-minute intervals during which the flask is removed from the bath, swirled vigorously, and returned to the bath. The reaction mixture should change color from colorless to deep red as the reaction progresses.

After completing the heating period, take the flask from the hot-water bath, remove the rubber stopper, and chill the flask in an ice bath. Add 50 mL of cold water, stopper tightly, and shake the flask vigorously to hydrolyze the polyphosphoric acid-product complex and any excess polyphosphoric acid. Disappearance of the red color may be used as an indicator of the completeness of the hydrolysis of the complex. Chill the flask thoroughly to complete precipitation of the product. Collect the crude product by suction filtration on the Hirsch funnel; wash with 5 mL of cold water, 5 mL of 10% aqueous sodium bicarbonate, and a second 5 mL of cold water; draw air through the filter until the product is reasonably dry.

Recrystallize the product from methanol (using decolorizing charcoal if the product is colored). Dry the product and determine its melting point. Approximate yield, 1.5 to 2 g; mp 75°.

Note 1   The lower aliphatic acids (butyric through capric) are liquids or low-melting solids that have distinctly unpleasant odors. Phenylacetic acid is also an aliphatic acid with an aromatic substituent (phenyl) on the α-carbon atom. It, too, has a very unpleasant odor. Therefore, take extreme care to keep these materials off the skin, clothing, books, and the laboratory bench. Use plastic gloves of the throw-away type in working with these acids. Furthermore, immediately wash after use or disassembly all glassware coming into contact with these acids. Very dilute (5%) aqueous sodium hydroxide followed by copious quantities of water or, for gummy or tarry residues, acetone followed by water will remove these materials very effectively from glassware or the laboratory bench. If any of these acids come in contact with the skin, wash them off immediately with soap and water. The skin absorbs all of these substances to some extent and you cannot completely wash them off with soap and water. If you employ a good scrubbing, any residual odor can be masked fairly well for several hours with a good spray deodorant. Usually the last traces of odor will be gone in 2–3 days to the great relief of family and friends.

Note 2   A metal boiling-water bath is preferable from the standpoint of safety. If a metal bath is not available, a 400-mL beaker half-filled with water and heated by a burner or, preferably, and electric hot plate will suffice.

B.   Preparation of *p*-Methoxybenzophenone

Caution!   *See **Caution** note in Part A of this experiment, page 318, concerning the handling of polyphosphoric acid.*

In a 125-mL flask place 20 g of polyphosphoric acid, 1.6 g of anisole, and 2.4 g of benzoic acid. Stopper the flask *loosely* with a rubber stopper and support it with a clamp in a boiling-water bath. After a minute or two, remove the flask and swirl to thoroughly mix the reactants. Return the flask to the water bath and continue heating for a period of one hour, removing the flask every 10 minutes or so to remix the contents by swirling. The reaction mixture should change color from colorless to amber.

After the heating period is complete, take the flask from the water bath, remove the rubber stopper, and chill the flask in an ice bath. Add 50 mL of cold water, stopper loosely, and swirl the flask vigorously to hydrolyze the polyphosphoric acid-product complex and any excess polyphosphoric acid. Disappearance of the amber color may be used as an indicator of the completeness of the hydrolysis of the complex. Chill the flask thoroughly to complete precipitation of the product. Extract the crude product, which has the form of a waxy solid, by adding to the flask 25 mL of ether. You may hasten the solution of solid material by the ether by warming in a water bath and swirling. Transfer the entire contents of the flask to a separatory funnel and separate the ether layer. Discard the lower aqueous layer. Wash the ether layer with a 50-mL portion of 10% aqueous sodium hydroxide solution and then with 50 mL of water. Discard both the basic and water washes. Transfer the ether layer to an evaporating dish. (**Caution!** *Ether is very flammable. Carry out the following step in a good fume hood. Extinguish all flames in the vicinity. Do NOT evaporate ether from a hot plate. [Note 1]*) Evaporate the ether from a steam bath or hot-water bath. The syrupy liquid residue will solidify on cooling. Yield, approximately 2.5 g. The crude ketone may be recrystallized from methanol, mp 62°.

Note 1   Although small volumes of relatively innocuous, volatile solvents may be evaporated in the hood, if facilities permit, it is best to remove solvents by simple distillation. Thus, in the present experiment, transfer the ether solution to the round-bottomed flask of a simple distillation setup. Add a Boileezer to the flask, and check the setup to make certain that the joints are tightly sealed. Cool the receiver in an ice bath. While there are no flames in the vicinity, distill off the ether using a filled metal hot-water bath. You may heat the water in a remote location in an ordinary kettle and transfer it to the water bath. One filling of hot water should distill about 100 mL of ether. Be prepared to remove the hot water if the distillation begins to proceed too rapidly.

REPORT 31

Name Section Date

Friedel-Crafts Reaction: Preparation of *p*-Anisyl Benzyl Ketone

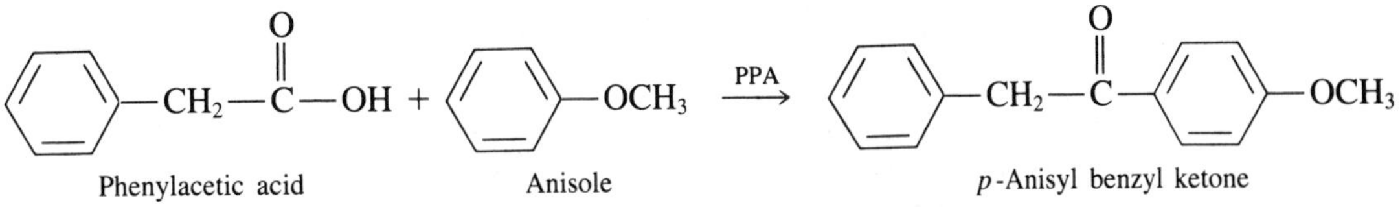

	Phenylacetic acid	Anisole	*p*-Anisyl benzyl ketone
Quantities	1.4 g	1.2 g	______ g
Mol. Wt.	______	______	______
Moles	______	______	______
Equivalents	______	______	______

Theoretical yield ______ g

Actual yield ______ g

Percentage yield ______

Preparation of *p*-Methoxybenzophenone

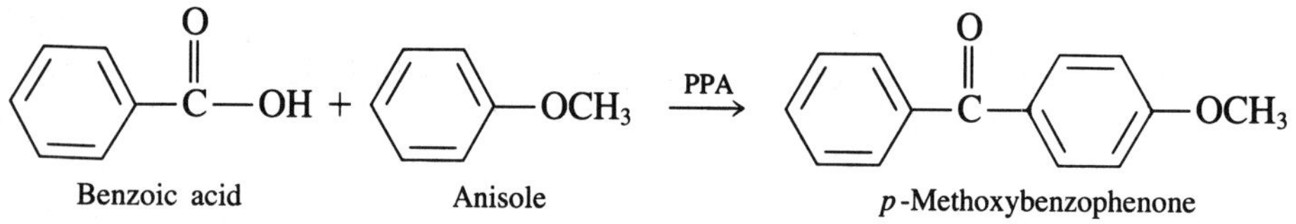

	Benzoic acid	Anisole	*p*-Methoxybenzophenone
Quantities	2.4 g	1.6 g	_____ g
Mol. Wt.	_____	_____	_____
Moles	_____	_____	_____
Equivalents	_____	_____	_____

Theoretical yield _____ g

Actual yield _____ g

Percentage yield _____

QUESTIONS and EXERCISES

1. What is the function of the Lewis acid catalyst in the Friedel-Crafts acylation reaction?

2. In the Friedel-Crafts reaction only catalytic amounts of catalyst are required for alkylation; however, a full molar equivalent or more of catalyst is required for acylation. Suggest an explanation.

3. What sequence of reactions would you follow in order to prepare 3-nitro-acetophenone via a Friedel-Crafts reaction?

4. Why does the *p*-isomer predominate in the product of Friedel-Crafts acylations?

5. If *p*-anisyl benzyl ketone were treated with a nitrating mixture, what product would most likely result?

6. What are some of the limitations of the Friedel-Crafts reaction?

7. Why is nitrobenzene sometimes used as a solvent for the Friedel-Crafts reaction?

A Molecular Rearrangement:
The Preparation of
Benzanilide from Benzophenone

An important class of molecular rearrangements is that in which an alkyl or aryl group with its pair of bonding electrons migrates to an electron-deficient carbon, oxygen, or nitrogen atom. Such rearrangements to nitrogen are usually irreversible and often are stereospecific.

The rearrangement of ketoximes when catalyzed by acids is a classic example of a molecular rearrangement. It is called the **Beckmann rearrangement** after the German chemist, who in 1866 was the first to discover that the oximes of certain carbonyl compounds rearranged to amides when treated with acids.

$$\underset{Ar'}{\overset{Ar}{>}}C{=}O + H_2NOH \cdot HCl \longrightarrow \underset{Ar'}{\overset{Ar}{>}}C{=}N\overset{OH}{\diagup} + H_2O$$

$$\underset{Ar'}{\overset{Ar}{>}}C{=}N\overset{OH}{\diagup} + H^+ \longrightarrow \underset{Ar'}{\overset{Ar}{>}}C{=}N\overset{\overset{+}{O}H_2}{\diagdown} \longrightarrow Ar{-}C{\equiv}\overset{+}{N}{-}Ar' + H_2O$$

$$Ar{-}C{\equiv}\overset{+}{N}{-}Ar' + H_2O \longrightarrow Ar{-}\overset{OH}{\overset{|}{C}}{=}N{-}Ar' \rightleftharpoons Ar{-}\overset{O}{\overset{\|}{C}}{-}\overset{H}{\overset{|}{N}}{-}Ar' + H^+$$

The mechanism usually proposed for the Beckmann rearrangement is that of a concerted reaction in which the migration of one group is accompanied by the departure of another.

An inspection of the reaction equation above also will show that the carbon-nitrogen double bond has the necessary geometry to permit the formation of two different stable oximes from an unsymmetrical ketone. Such isomers are formed by *syn* and *anti* modes of addition and are related respectively to the *cis-* and *trans-* forms of isomers first encountered in doubly bonded carbon-carbon structures. Regardless of the oxime isomer formed, the migrating group is always the one that is *anti* to the hydroxyl group. The orientation of the groups in the starting oxime, therefore, may be determined from an examination of the resultant amide.

Beckmann used phosphorus pentachloride to effect the rearrangement of the oximes in his original investigation. However, the rearrangement of oximes may be accomplished by any one of a number of acids. In the present experiment we will illustrate the facility with which the rearrangement may be accomplished with polyphosphoric acid.

PROCEDURE

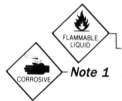

*Read the **Caution** note given in Experiment 31 regarding the handling of polyphosphoric acid before beginning this experiment.*

In a 125-mL Erlenmeyer flask place 20 g of polyphosphoric acid, 1.82 g of benzophenone, and 2.1 g of hydroxylamine hydrochloride. Place the flask in a boiling-water bath and for a minute or two rotate and swirl in order to thoroughly mix the reactants. Clamp the flask in position and continue heating for 30–45 minutes or until frothing has ceased. (Note 1) At the end of the reaction period add 50–75 g of crushed ice to the reaction mixture and shake. Benzanilide will begin to precipitate almost immediately. After all the ice has melted, collect your product on the Büchner funnel. Wash twice with 10-mL portions of ice water, dry, and weigh. Yield, 1.5–1.7 g; mp 161°. The benzanilide will be reasonably pure as collected but may be recrystallized from hot ethanol.

Note 1 Because hydrogen chloride evolves, it is best to run this reaction in a good fume hood. However, if this is not possible, fit the Erlenmeyer flask with an acid vapor trap such as that described in Experiment 35(a), Part I, and in Figure 35.1. The reaction may be run in a round-bottomed flask to facilitate the attachment of the acid vapor trap.

The Preparation of Benzanilide from Benzophenone

Reaction equation

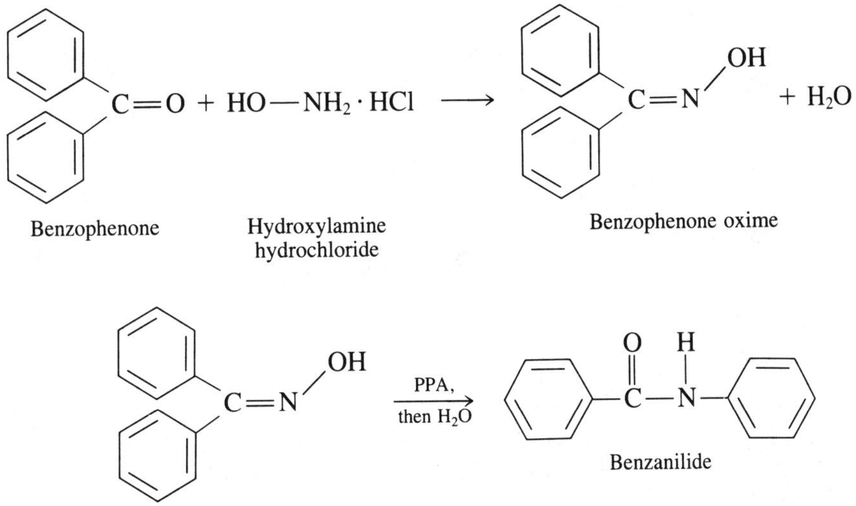

| | Benzophenone | Hydroxylamine hydrochloride | Benzophenone oxime |

Benzanilide

	Ketone	HONH$_2$ · HCl	Anilide
Quantities	1.82 g	2.1 g	_____
Mol. Wt.	_____	_____	_____
Moles	_____	_____	_____
Equivalents	_____	_____	_____

Theoretical yield _____ g

Actual yield _____ g

Percentage yield _____

mp _____ °C

1. Outline a procedure for the preparation and isolation of aniline from benzo-phenone.

2. What products would be possible from a Beckmann rearrangement of ace-tophenone?

3. Aliphatic aldehydes and ketones do not give isomeric oximes. Suggest a reason for this.

4. If the oxime of cyclohexanone were treated with acid, what would be the structure and name of the rearranged product?

5. Nylon 6,
$$-N\left[\begin{array}{c}H\\|\\N\end{array}-\begin{array}{c}O\\||\\C\end{array}-(CH_2)_4-\begin{array}{c}O\\||\\C\end{array}-\begin{array}{c}H\\|\\N\end{array}-(CH_2)_6-\begin{array}{c}H\\|\\N\end{array}\right]_n$$
, may be manufac-tured from cyclohexanol. One of the necessary steps in the process is a Beckmann rearrangement. Write a series of reactions that would lead to the production of this linear polymer.

Polymerization

Polymerization is a reaction in which many single molecules (monomers) are united to form giant molecules or **polymers.** The latter are high molecular weight compounds that often possess properties that make them very useful as structural materials. Reactions leading to the production of polymers are of two principal types. If the polymer is formed by a chain reaction of the monomer, the reaction is referred to as **chain,** or **addition, polymerization.** Common examples of polymers formed in this manner are polyethelene $+CH_2CH_2\rightarrow_n$, Teflon, $+CF_2CF_2\rightarrow_n$, and many of

the very useful vinyl polymers, $+CH_2-\overset{\overset{\displaystyle H}{|}}{\underset{\underset{\displaystyle X}{|}}{C}}\rightarrow_n$. Chain polymerization often pro-

ceeds via a *free-radical* chain reaction mechanism catalyzed by a peroxide, ultraviolet light, or some agent that aids in the formation of a free radical. The reaction involves three different steps: (1) initiation, (2) propagation, and (3) termination.

Initiation:
$$CH_2{=}CH_2 + R\cdot \longrightarrow R{-}CH_2{-}CH_2\cdot$$

Propagation:
$$RCH_2{-}CH_2\cdot + n\,CH_2{=}CH_2 \longrightarrow R{+}CH_2{-}CH_2\rightarrow_n CH_2CH_2\cdot$$

Termination:
$$R{+}CH_2{-}CH_2\rightarrow_n CH_2CH_2\cdot + R\cdot \longrightarrow R{+}CH_2CH_2\rightarrow_{n+1}R$$

When the amount of initiator (R, the free radical) is small and n is a large number (1000–2000), the molecular weight of the polymer becomes very great. Another type of polymerization reaction leading to the formation of high molecular weight compounds is **step,** or **condensation, polymerization.** Interaction of functional groups in this type of polymerization reaction takes place in a stepwise fashion, usually producing small molecules (water, hydrogen chloride, ammonia) as byproducts in each step. Cellulose and proteins are examples of natural polymers of this type. Restoring the simple molecules (water, in this case) by hydrolysis of these natural substances results in the formation of the unit molecules from which they were formed (that is, simple sugars and α-amino acids, respectively). Examples of step polymers familiar to almost everyone are the synthetic fibers Nylon (a polyamide) and Dacron (a polyester).

There are a number of ways in which a step polymer such as Nylon can be prepared; however, the procedure that is most suitable for the beginning laboratory student is based on the reaction of a primary amine with an acyl chloride to yield an amide. Thus, the reaction of a diacid chloride with a diamine can proceed under the proper conditions to form a polyamide in which the amide linkage, $-\overset{\overset{\text{O}}{\|}}{\text{C}}-\overset{\overset{\text{H}}{|}}{\text{N}}-$, recurs repeatedly to produce a linear polymer of approximately 12,000 molecular weight. Nylon 6-6 and Nylon 6-10 are examples of such linear polyamides, in which the 6-6 and 6-10 designations refer to the 6-carbon diamine and the 6- and 10-carbon diacids, respectively, from which the polymers originate. In Procedure B of the present experiment Nylon 6-10 is prepared from sebacoyl chloride and hexamethylenediamine by a procedure that is sometimes called "The Nylon Rope Trick."

$$n \underset{\text{Cl}}{\overset{\text{O}}{\diagdown}}\text{C}-(\text{CH}_2)_8-\text{C}\underset{\text{Cl}}{\overset{\text{O}}{\diagup}} \;+\; n \underset{\text{H}}{\overset{\text{H}}{\diagdown}}\text{N}-(\text{CH}_2)_6-\text{N}\underset{\text{H}}{\overset{\text{H}}{\diagup}} \longrightarrow$$

Sebacoyl chloride   Hexamethylenediamine

$$\text{Cl}\left[\overset{\overset{\text{O}}{\|}}{\text{C}}-(\text{CH}_2)_8-\overset{\overset{\text{O}}{\|}}{\text{C}}-\overset{\overset{\text{H}}{|}}{\text{N}}-(\text{CH}_2)_6-\overset{\overset{\text{H}}{|}}{\text{N}}\right]_n \text{H} \;+\; n\;\text{HCl}$$

Nylon 6-10

The chemical industry devotes much of its research activity and its manufacturing facilities to the production of polymers, yet polymerization reactions as laboratory exercises seldom are considered in the same perspective as the more classical preparations. There are several reasons for a lack of emphasis on polymerization reactions in the laboratory program. The reactions involved in the formation of a polymer in many cases are very complex and not as illustrative of organic reactions as are simpler preparations, and starting materials required usually are not easily available. The laboratory preparation of a polyurethane foam, while representing no exception to these shortcomings, illustrates a polymerization reaction of great commercial importance. The reaction is easy to carry out, rather spectacular to observe, and can be adapted to a very practical purpose.

In Procedure A of this experiment a polyurethane foam is prepared from two substances of natural origin, castor oil and glycerol. Although commercial foams are not likely to be prepared from such simple starting materials, the principles behind the laboratory and commercial preparations are the same.

The simple urethanes may be prepared by the reaction of alcohols with isocyanates according to the following equation.[1]

---

[1] Although no small molecule is liberated during polymer formation, these polymers are labeled condensation polymers because the polymer *could be considered* to result from the reaction of a carbamic acid and an alcohol with the loss of water.

$$R-N=C=O + R'OH \longrightarrow R-\overset{\overset{\displaystyle H}{\displaystyle |}}{N}-\overset{\overset{\displaystyle O}{\displaystyle \|}}{C}-OR'$$

An	An	A urethane
isocyanate	alcohol	

The polyurethanes are prepared from diisocyanates and polyhydric alcohols. Reactants with two or more functional groups are capable of polymerizing until all material is incorporated into one giant, cross-linked molecule. A small amount of water in the reaction mixture results in the production of sufficient unstable carbamic acids which, on decomposition, provide the carbon dioxide that causes the foam to rise.

$$R-N=C=O + H_2O \longrightarrow R-\overset{\overset{\displaystyle H}{\displaystyle |}}{N}-\overset{\overset{\displaystyle O}{\displaystyle \|}}{C}-OH$$

A substituted
carbamic acid

$$R-\overset{\overset{\displaystyle H}{\displaystyle |}}{N}-\overset{\overset{\displaystyle O}{\displaystyle \|}}{C}-OH \longrightarrow R-NH_2 + CO_2$$

The process whereby a polyurethane foam is produced is somewhat like baking a cake. The small carbon dioxide bubbles fill the polymeric material with innumerable pores that make the foam "light." If these pores are closed the foam is a rigid one and the enmeshed gas is trapped to provide the remarkable insulating and buoyant properties for which this type of material is noted. The structural unit of a polyurethane is shown on the following page.

$$2n \quad \underset{\substack{\text{4-Methyl-}m\text{-phenylene} \\ \text{diisocyanate} \\ \text{(Tolyl diisocyanate, TDI)}}}{\text{[ring with CH}_3\text{, N=C=O, N=C=O]}} + 2n \quad HO(CH_2)_x-\overset{\overset{\displaystyle OH}{\overset{\displaystyle |}{(CH_2)_y}}}{\underset{\displaystyle H}{\overset{\displaystyle |}{C}}}-(CH_2)_z-OH \longrightarrow$$

A polyhydric alcohol[1]

_____

[1] The molecular weight of a polyhydric alcohol used in a polyurethane foam may vary from 500 to 3000. The small letters x, y, and z therefore represent large numbers (10–70). While the alcohol in the sample equation is shown as a simple triol, commercial Polyols contain additional hydroxyl groups as well as ether linkages.

$$\left[\begin{array}{c} \underset{\displaystyle \text{CH}_3}{\bigcirc}\!-\!\underset{\displaystyle \underset{\displaystyle \text{N}-\text{H}}{|}}{\overset{\displaystyle \text{H}}{\text{N}}}\!-\!\overset{\displaystyle \overset{\displaystyle \text{O}}{\|}}{\text{C}}\!-\!\text{O}\!-\!(\text{CH}_2)_x\!-\!\overset{\displaystyle \text{H}}{\underset{\displaystyle (\text{CH}_2)_y}{\text{C}}}\!-\!(\text{CH}_2)_z\!-\!\text{O} \end{array} \right]_n$$

Structural unit of a polyurethane

This is a "fun" experiment; however, some of the chemicals involved are now thought to be toxic, and it may be that your instructor will prefer to carry out this experiment for you as a laboratory demonstration.

PROCEDURE

Caution! Several of the chemicals used in this experiment are toxic and/or corrosive; therefore, you should wear disposable plastic gloves and a plastic apron. Tolulene 2,4-diisocyanate (TDI) is a highly toxic lachyrmator. Sebacoyl chloride is a corrosive lachyrmator. 1,6-Hexanediamine (hexamethylene diamine) is corrosive. Avoid contact with these materials. Wash off any spills with liberal use of soap and water.

A. The Preparation of a Polyurethane Foam

In a plastic lined or waxed container (Note 1) weigh out 35 g of castor oil and 10 g of glycerol. To the combined reactants add 5 drops of stannous octoate, 10 drops of silicone oil (Dow-Corning 200), and 10 drops of water (use a dropper). Blend these ingredients thoroughly by stirring with a glass rod to a creamy, viscous mass. Next, weigh in a separate container (paper cup or beaker) 30 g of 4-methyl-*m*-phenylene diisocyanate (tolulene diisocyanate, TDI). (Note 2) Add the correctly weighed amount of TDI *in one portion* to the previously mixed ingredients. Mix rapidly and thoroughly to a smooth, creamy, homogeneous mixture. When the container becomes warm to the touch and small bubbles begin to form, immediately discontinue stirring, place the container on a piece of newspaper *in a fume hood*, and allow the reaction to proceed spontaneously. After foaming has ceased, allow the polymeric material to cool and set for 4 or 5 hours before removing the form. The volume of foam produced by the above formulation is approximately 75 cubic inches. (Note 3) When the foam is fairly firm show it to your instructor, who may ask you to slice off a small section with a sharp knife to attach to your report.

Polyurethane foams produced from castor oil and glycerol usually will be of the compressible type and will set with varying degrees of shrinkage. Although the practical applications of such foams are limited, the reaction is an excellent one to illustrate the polyfunctional requirements for the preparation of this type of elastometer.

Note 1   A paper milk carton (quart capacity) with the top removed or a paper cup (milkshake size) serves very well as a container for the foaming reaction.

Note 2   TDI reacts vigorously with the moisture in air or in the skin to ultimately produce substituted ureas. Recap bottle promptly after using.

Note 3   Freshly formed foam often is sticky to the touch. Should you accidentally get any newly formed foam on the workbench or hands, remove it with a 1 : 1 isopropyl alcohol-acetone mixture.

B. Preparation of Nylon 6-10

Caution!   Wear disposable plastic gloves while doing this part of the experiment, for it is very awkward to handle the Nylon polymer without some contact with the hands.

Examine Figure 33.1 and gather the apparatus needed: ring-stand, clamp, a 15–20 × 60 mm vial (a litmus paper or pHydrion paper vial will do nicely), and a 6-in length of copper wire with one end bent into a small hook. Using a cotton swab or your little finger, coat the inner surface of the vial with a *thin* layer of silicone stopcock grease or silicone oil. (Note 1) Fill the vial to about the 40% mark with a 5% solution of sebacoyl chloride in methylene chloride. Hold the vial in an inclined position and very slowly add onto the methylene chloride layer an equivalent volume of a 5% aqueous solution of hexamethylenediamine (1,6-hexanediamine). A reaction product should form immediately at the interface of the two immiscible layers. Clamp the vial to the ring-stand. Reach through the upper layer with your hook and draw up the Nylon that has formed (see Fig. 33.1). This will expose fresh reactants from both layers to produce additional polymer in the form of a hollow tube. Slowly and steadily continue to remove new material as one continuous hollow thread. Wind the thread onto a piece of cardboard or a cardboard tube as it forms or lead it into a 600-mL beaker filled with water. If the thread breaks, use the hook to start a new thread. When the reactants have been reduced to about one-half to one-third of their initial volume, stopper the vial and shake vigorously to thoroughly mix the remainder of the two layers. Transfer the resulting white opaque mass to a beaker and wash once with 95% ethyl alcohol and once with water. Squeeze the lump of material as

Figure 33.1   Removing the Nylon thread from the interface of two immiscible layers.

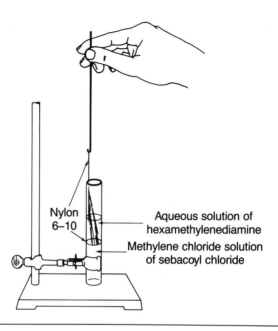

Nylon
6–10

Aqueous solution of
hexamethylenediamine

Methylene chloride solution
of sebacoyl chloride

dry as possible between paper towels and examine it. How easy is it to pull apart; that is, does it have any tensile strength in this form? Transfer the mass to a small test tube or to a metal spoon. Try to melt the polymer gently over a low flame or on the hot plate. Do not heat it much above its melting point or it will darken and char. Does it melt easily? If you succeed in melting it, try drawing a thin fiber from the melt with your wire hook. A fiber 1 to 2 cm in length is long enough. If you are successful in drawing a fiber from the molten material, compare its strength with that of the thread produced from the two layers. Attach a short length of your "Nylon rope" to your report form.

Note 1   Coating the inner surface of the sample vial prevents the Nylon from sticking to the glass wall and ensures a continuous thread.

Name   Section   Date

Polymerization

Attach a thin section of your polyurethane foam below.

Attach a short length of your "Nylon rope" below.

1. Write the equations for the preparation of the following polymers: (a) Dacron, (b) Orlon, (c) Teflon, (d) polystyrene.

2. Define the term *copolymer*. Give an example (structural formula) of a copolymer.

3. The principal difference between *chain* and *step* polymerization is that in chain polymerization all of the intermediates are unstable species such as radicals, anions, or cations, whereas in step polymerization the intermediates are reasonably stable organic molecules. Illustrate this difference by writing the equations for the reactions of three molecules of ethylene by a free-radical chain process and equations for the reactions of three molecules of sebacoyl chloride with three molecules of hexamethylenediamine. Until very recently, these processes were called *addition* and *condensation* polymerization, respectively. Why do you suppose polymer chemists chose the new names above?

EXPERIMENT

34

Time: $3\frac{1}{2}$–4 hours

The Coupling of Aromatic Diazonium Compounds; Dyes and Dyeing

Nearly one-half of all synthetic dyes now in use are azo dyes prepared by coupling aromatic diazonium salts with phenols, naphthols, or aromatic amines.

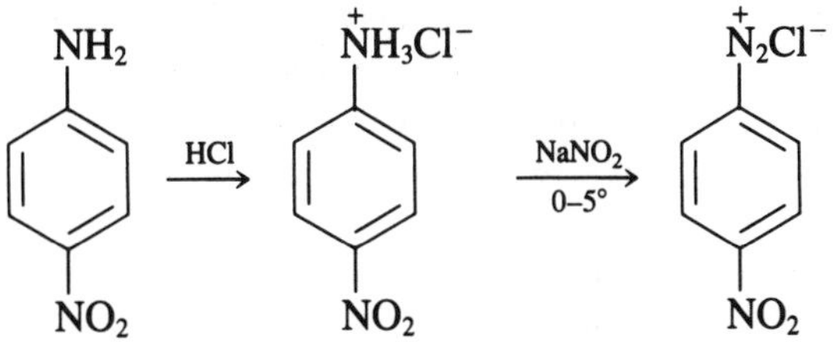

p-Nitrobenzene-diazonium chloride

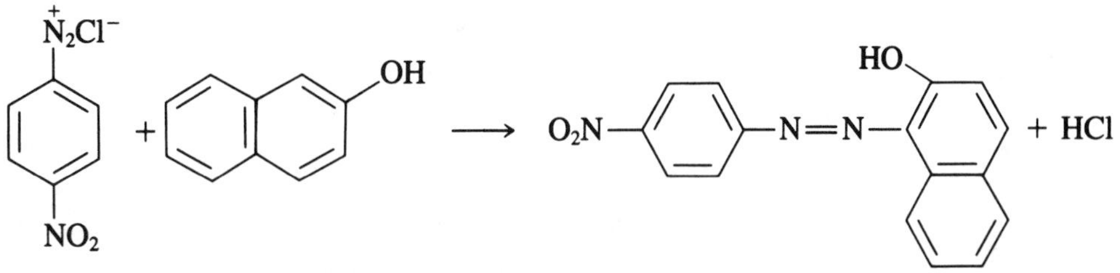

β-Naphthol　　　Para red

The coupling reaction is carried out in alkaline, neutral, or in weakly acidic solution. Inasmuch as the diazotization and coupling reactions take place within the fabric at the temperature of an ice bath, azo dyes produced in this manner sometimes are referred to as "ice colors." One of the first dyes of this type to be prepared was **para red,** the preparation of which is part of the present experiment. Its use illustrates a method of dyeing known as **ingrain** dyeing. Two other common methods of dyeing—**mordant** and **direct**—also will be illustrated.

Precautions. *This experiment requires the use of hydrochloric acid, sodium hydroxide,* p*-nitroaniline, 2-naphthol, potassium antimonyl tartrate, picric acid, and tannic acid. See HAZARD CATEGORIES 2, 3, 4, 6, and 8, page 4.*

PROCEDURE

Caution! Although the procedures and methods used in this experiment have been used for many years in the dye industry, it is now known that many of the chemicals used are toxic or cancer suspect. The following are known to be toxic: malachite green, methyl violet, potassium antimony tartrate, picric acid, and p-nitroaniline (very toxic). 2-Naphthol (β-naphthol) is an irritant, and tannic acid is cancer suspect. In addition, some of the dyes prepared will dye your skin or your clothing. It is very important that you exercise extreme care in dyeing operations. Avoid handling samples with the fingers; use sturdy glass rods or metal tongs. Avoid splashing or spilling. Wear disposable plastic gloves and a plastic apron or some other protective garment.

A. Azo Dyes and Ingrain Dyeing (Para Red)

Place 1.0 g of *p*-nitroaniline in a 50-mL beaker; add 25 mL of water and 1 mL of concentrated hydrochloric acid. Place the beaker in an ice bath and cool to 0–5°. Slowly stir into the cold *p*-nitroaniline hydrochloride solution a cold (0–5°) solution of 1.0 g of sodium nitrite in 5 mL of water. While the diazonium salt solution is in the ice bath, prepare in a 250-mL beaker a solution of 1 g of β-naphthol in 100 mL of 2% sodium hydroxide. Immerse a piece of clean, cotton cloth or bandage gauze (1 × 2 inches) in the alkaline β-naphthol solution, squeeze out as much solution as possible, and hang the cloth up to dry. Dilute the cold diazonium salt solution with 200 mL of ice water and dip the dry, β-naphthol-soaked cloth in the solution of diazotized amine. Puddle the cloth with a glass rod, remove, and again air dry. Attach the dry, dyed cloth to your report form.

B. Triphenylmethane Dyes; Mordant Dyeing

Prepare the following solutions.

1. Two dye baths, prepared from (a) 0.1 g malachite green in 200 mL of hot water and (b) 0.1 g methyl violet in 200 mL of hot water.

2. A mordant bath prepared by dissolving 0.2 g of tannic acid in 150 mL of water.

3. A fixing bath prepared by dissolving 0.2 g tartar emetic (potassium antimonyl tartrate, $KSbOC_4H_4O_6$) in 150 mL of water.

Each of these solutions must be heated to 80–90° on the steam bath, hot plate, or over a burner flame before use.

Soak four clean pieces of cotton cloth in the hot (80–90°) mordanting solution for one minute, squeeze as dry as possible, and transfer two pieces to the fixing bath. Allow the samples to remain in the fixing bath at 80–90° for 5 minutes. Remove the cloth samples from the fixing bath and transfer to the two separate dye baths. Allow each of the cloth samples to remain in the hot (90°) dye baths for about 5 minutes. Remove, wash with cold water, and hang them up to dry. Repeat the dyeing process with the two pieces of cotton cloth that have had the tannic acid treatment but have not had a fixing bath. Repeat the dyeing process using two pieces of cotton cloth that have had neither previous treatment. Which process produces a fast color? Attach all six samples to your report form.

C.   Nitro Dyes; Direct Dyeing

Dissolve about 0.1 g of trinitrophenol (picric acid) (**Caution!**   *Flammable solid.*) in 250 mL of hot water. Immerse in the bath pieces of wool, silk, and cotton for 2–3 minutes. Remove the cloth, rinse, and dry. Attach the samples to your report form.

Name   Section   Date

Azo Dyes and Ingrain Dyeing (Para Red)

> Cotton

Triphenylmethane Dyes and Mordant Dyeing

> Cotton
> (dye only)

> Cotton (mordant
> and dye)

> Cotton (mordant,
> fix, and dye)

Nitro Dyes and Direct Dyeing

> Cotton

> Wool

> Silk

1. What substances other than tannic acid may be used as mordants?

2. Why is picric acid a good direct dye for silk and wool but not for cotton?

3. With the aid of your text draw the structures of (a) indigo, (b) Congo Red, and (c) malachite green. Enclose the chromophore with a broken line. Use a solid line to encircle the auxochromes in (a) and (c).

4. Write the equation for the probable method of synthesis of Congo Red (diazotization and coupling steps only).

5. In the red form, Congo Red may be considered to be a salt in which the —SO_3H groups have been converted to —SO_3^- groups (by treatment with base). In the blue form, two protons are added to the molecule to give two substances (differing principally in the location of the protons) in equilibrium with each other. One of these has the protons added to the —NH_2 groups to form —NH_3^+ groups. What is the probable site of protonation in the other substance (which is largely responsible for the blue color)? (*Hint:* It is *not* the SO_3^- group!) Write two resonance structures for the second protonated substance that will show the nature of the chromophore responsible for the blue color.

A Reaction Sequence: The Preparation of 4-*tert*-Butylbenzoic Acid

This preparation illustrates two useful and general reactions: (1) the bromination of aromatic hydrocarbons, and (2) the carbonation of the Grignard reagent. It also gives the student opportunity to carry out a short reaction sequence. The sequence involves four successive steps as indicated by the following equations: (1) The preparation of 4-bromo-*tert*-butylbenzene, (2) the formation of 4-*tert*-butylphenylmagnesium bromide, (3) the carbonation of the Grignard reagent, and (4) the hydrolysis of the Grignard addition compound.

(1) [structure: C(CH₃)₃-substituted benzene] + Br₂ $\xrightarrow{\text{Fe}}$ [structure: 4-bromo-tert-butylbenzene with C(CH₃)₃ and Br]

(2) [structure: C(CH₃)₃-substituted bromobenzene with Br] + Mg $\xrightarrow[\text{ether}]{\text{Anhydrous}}$ [structure: C(CH₃)₃-substituted benzene with MgBr]

(3) [structure: benzene with MgBr and C(CH₃)₃] + CO₂ $\longrightarrow$ [structure: benzene with C(=O)—O—MgBr and C(CH₃)₃]

$$\text{(4)} \quad \underset{\underset{\displaystyle \text{C(CH}_3)_3}{\big|}}{\overset{\overset{\displaystyle \text{O}}{\|}}{\text{C}-\text{O}-\text{MgBr}}} \quad \xrightarrow{\text{H}_3\text{O}^+} \quad \underset{\underset{\displaystyle \text{C(CH}_3)_3}{\big|}}{\overset{\overset{\displaystyle \text{O}}{\|}}{\text{C}-\text{OH}}}$$

The continuity of the sequence may be interrupted after step (1) or (3). Once the Grignard reagent is formed, it should be carbonated without delay. After the Grignard reagent has been carbonated, it may stand until the next laboratory period.

> **Precautions.** *This experiment requires the use of ether, bromine, hydrochloric acid, and sodium hydroxide. See HAZARD CATEGORIES 1, 2, and 3, page 4.*

PART I Preparation of 4-Bromo-*tert*-Butylbenzene

PROCEDURE **Caution!** *Hydrogen bromide, which is evolved during this experiment, is very corrosive to the skin and mucous membranes. Avoid contact with and breathing of its vapors. Unless the experiment is carried out in a good exhaust hood, the apparatus must be fitted with an acid vapor trap as described below.*

Assemble an apparatus consisting of a clean, dry 250-mL round-bottomed flask and a reflux condenser fitted with one of the gas (hydrogen bromide) absorption devices shown in Figure 35.1. One product of the reaction is hydrogen bromide, an acidic gas, and unless you carry out the reaction in a fume hood, the gas must be absorbed. After your instructor has approved your apparatus, remove the reaction flask and place in it 26.8 g (0.2 mole, 31 mL) of *tert*-butylbenzene. Add 24 g (0.15 mole, 8.2 mL) of bromine from the bromine buret (**Hood!**) directly to the *tert*-butylbenzene. Refit the condenser and circulate water *slowly* through the water jacket. Have ready a pan of crushed ice and water to arrest the reaction should it become too vigorous. Through the open top of the condenser introduce five #3 tacks. (Note 1) Connect the gas trap to the condenser. Reaction frequently is immediate, as the rising of hydrogen bromide bubbles through the surface of the mixture indicates. Should the reaction become too lively, immerse the flask in the ice water without delay. The reaction, if too rapid, will cause bromine vapor to be carried over into the gas trap. The reaction should continue at a satisfactory rate without the necessity of continued cooling and may be considered complete when the space above the surface of the reaction mixture is clear of bromine vapor and bubbling at

Figure 35.1 Devices for the absorption or removal of acidic gases during refluxing.

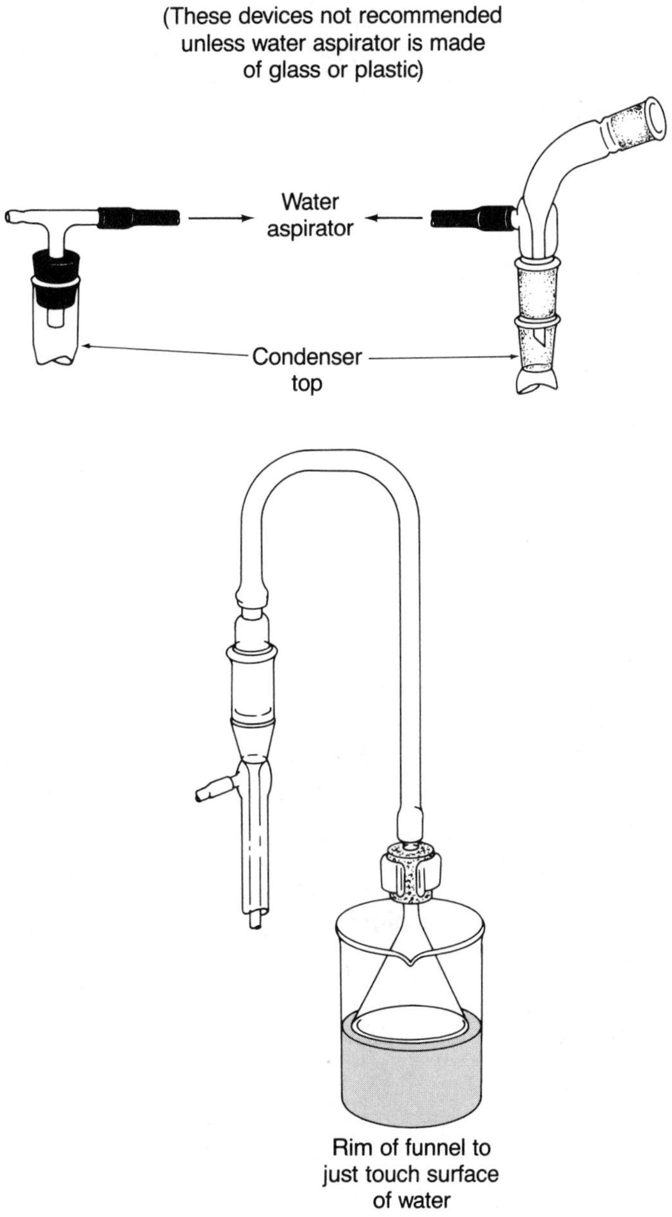

(These devices not recommended
unless water aspirator is made
of glass or plastic)

Water
aspirator

Condenser
top

Rim of funnel to
just touch surface
of water

the surface ceases (approximately 45 minutes). Heat the reaction mixture in a warm-water bath (75°) for 10–15 minutes. Remove the gas trap and through the top of the condenser add 75 mL of cold water. Remove the condenser, transfer the reaction mixture to a separatory funnel, and separate the layers. Discard the aqueous layer. Wash the organic layer a second time with 75 mL of 5% sodium carbonate solution. (**Caution!** *Effervescence.*) Separate and wash a third time with 50 mL of water. (Note 2) Draw off and discard the wash water. Dry the organic layer over calcium chloride. (Note 3) While your product is drying set up the apparatus required to carry out the procedure described under Part II. (Note 4)

Note 1   Old, used tacks serve as an excellent catalyst for this reaction. If you use new, blued tacks, clean them by immersing in dilute nitric acid for 1–2 minutes or file them to brightness.

Note 2   If your organic layer is orange or brown in color owing to the presence of unreacted bromine, add 10 mL of a saturated solution of sodium bisulfite in water to the 50 mL of water used for the third wash, and wash the organic layer a fourth time with another 50-mL portion of water.

Note 3   You may interrupt the sequence at this stage and perform optional Experiment 35(b) to determine the percentage yield.

Note 4   The high boiling point of *tert*-butylbenzene (169°) and that of 4-bromo-*tert*-butylbenzene (232°) preclude an easy separation of the reaction product by any means other than a fractional distillation under diminished pressure. The entire reaction mixture containing approximately 24 g (75% yield) of 4-bromo-*tert*-butyl-benzene is used in the preparation of the Grignard reagent. Any unreacted *tert*-butylbenzene will remain in the solvent ether. Should the instructor prefer a separation of 4-bromo-*tert*-butylbenzene, it may be distilled at 80–81°/2 mm. Yield, 75–78%.

PART II   The Preparation of a Grignard Reagent in the 4-*tert*-Butylbenzoic Acid Sequence

The preparation of 4-*tert*-butylphenylmagnesium bromide is accomplished by following a procedure only slightly modified from that described in Experiment 24 for the preparation of *n*-butylmagnesium bromide.

PROCEDURE   *Caution!   Exercise care in handling Dry Ice. Contact with the skin can cause severe burns or frostbite. Always use cotton gloves or tongs.*

Assemble an apparatus such as that illustrated in Figure 24.1 using clean, *dry* glassware and one ring-stand. Place 2.64 g (0.11 mole) of magnesium turnings in the flask, wet the turnings with 2–3 mL of 4-bromo-*tert*-butylbenzene, then add 15 mL of *anhydrous* ether, and a small (pinhead size) crystal of iodine in the flask. Into a dropping funnel place the remainder of the reaction mixture and an additional 35 mL of *anhydrous* ether. Swirl the mixture to ensure complete solution. Add at one time approximately half of the ether solution of 4-bromo-*tert*-butylbenzene to the magnesium turnings. Warm the mixture to a gentle reflux on a 50° water bath. You will know that the reaction has begun when the pale amber color of the solution fades and becomes an opalescent white and refluxing of the ether becomes spontaneous without further heating. Add the remainder of the reagent from the dropping funnel dropwise to maintain a fairly rapid reflux. Cooling of the refluxing mixture is unnecessary unless violent boiling and frothing occur and ether vapor escapes from the top of the condenser. After the reaction is well on its way, you may observe a darkening of the mixture. When refluxing subsides and nearly all of the magnesium has been consumed, immerse the reaction mixture in a 50° water bath and allow it to reflux for an additional 15 minutes. Remove the condenser from the flask and pour the entire contents slowly but steadily onto 35 g of crushed Dry Ice contained in a 600-mL beaker, taking care to keep unreacted magnesium in the flask. (Note 1) Cover the reaction mixture with a watch glass and let stand until the excess Dry Ice has completely sublimed. The Grignard addition compound will appear as a viscous glassy mass. If the mass is too viscous to stir, add an additional 50 mL of ether. (Note 2)

Hydrolyze the Grignard addition product by adding to it 50 g of crushed ice to which 15 mL of concentrated hydrochloric acid has been added. Stir the mixture until two layers appear, then transfer the mixture to a separatory funnel. Draw off the lower aqueous layer and discard it. Wash the upper ether layer once with 50 mL of water (Note 3); then, separate and discard the wash water layer. Extract the ether layer three times with 50-mL portions of 5% sodium hydroxide solution. Combine the basic extracts and discard the ether layer. Wash the combined basic extracts once with 50 mL of ether. Discard the ether layer. Place the combined basic extracts in a 400-mL beaker, set in a boiling-water bath (**Hood!** *Extinguish all flames!*), and stir until the ether dissolved in the alkaline solution has been boiled out. (Note 4) Cool

the alkaline solution in an ice-water bath and precipitate the 4-*tert*-butylbenzoic acid by the addition of 20 mL of concentrated hydrochloric acid. Collect the precipitated acid on a Büchner funnel by suction filtration. Wash the collected crystals with several small portions of cold water and dry. The product is fairly pure as collected, but a 0.5-g sample may be recrystallized from dilute alcohol for a melting point determination. Weigh the product and turn it in with your report. Yield, approximately 11 g (56%); mp 164°.

Note 1   The Dry Ice is best broken up by wrapping one lump in a clean, dry towel and beating it on the desk top or floor.

Note 2   You may interrupt the reaction sequence at this stage and complete it in the next laboratory period.

Note 3   If the ether layer appears dark yellow or brown due to the presence of iodine, add 10 mL of a saturated solution of sodium bisulfite in water to the 50 mL of wash water.

Note 4   Ether is soluble in water to the extent of 7%. Unless the ether is removed before the substituted benzoic acid is precipitated, the product may appear as a waxy solid and not crystalline.

Preparation of 4-*tert*-Butylbenzoic Acid

Reaction
equation

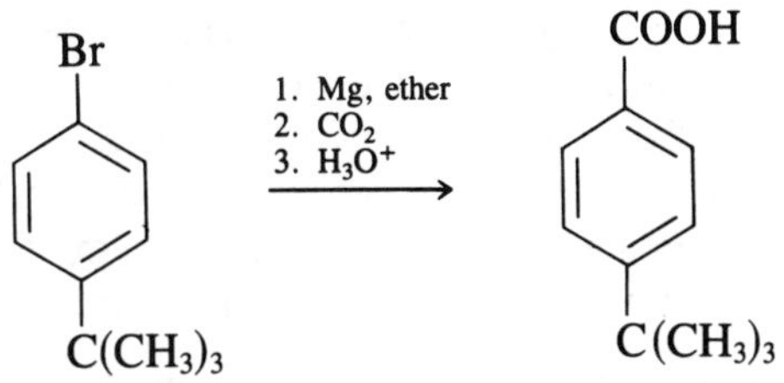

4-Bromo-*tert*-butylbenzene   4-*tert*-Butylbenzoic acid

Quantities   24 g*   _____

Mol. Wt.   _____   _____

Moles   _____   _____

Equivalents   _____   _____

Theoretical yield _____

Actual yield _____ g

Percentage yield _____

mp _____ °C

*Based on an assumed 75% yield of 4-bromo-*tert*-butylbenzene.

QUESTIONS
and
EXERCISES

1. What is the function of the iron tacks used in the preparation of 4-bromo-*tert*-butylbenzene? Write a mechanism for the bromination of *tert*-butylbenzene that supports your answer. What other catalysts could be substituted for the tacks?

2. What is the function of anhydrous ether in the preparation of a Grignard reagent? In some advanced work, in addition to the use of dry ether, a slow stream of dry nitrogen is passed through the vessel in which a Grignard reagent is being prepared or used. From what is the Grignard reagent being protected by the dry nitrogen? Write equations for two or three reactions that could take place to destroy all or part of the Grignard reagent if it were not protected.

3. Two easily available forms of carbon dioxide are Dry Ice and the compressed gas. Dry Ice has the obvious advantage of being easy to use in the carbonation of Grignard reagents. What is an obvious disadvantage of the use of this form of carbon dioxide? What can be done to minimize the incidence of this side reaction?

4. What products would you expect from the addition of phenylmagnesium bromide to (a) methyl ethyl ketone, (b) methyl acetate, and (c) acetonitrile?

(Optional) The Gas Chromatographic
Separation and Identification
of the Reaction Product Resulting
from the Bromination
of *tert*-Butylbenzene

An interesting departure from the sequence of steps comprising Experiment 35 is the gas chromatographic analysis of the reaction mixture that results from the bromination of *tert*-butylbenzene. An approximate percentage yield of 4-bromo-*tert*-butylbenzene may be determined via GLC and used as a basis for calculation of the final yield of 4-*tert*-butylbenzoic acid.

PROCEDURE
The procedure that follows is essentially the same as that described in Experiment 7, Part IV, except for the following changes in instrument adjustment. Set an optimum flow rate for the carrier gas (Note 1), and set a temperature of 235° for the injection port. Start the recorder and set the attenuation at a sensitivity value that ensures a good measureable signal. Draw a 0.5-μL sample of the unrectified but dried reaction mixture into the syringe and follow this liquid sample by drawing in an additional 5 μL of air. Insert the needle into the septum, being careful to avoid contact with the hot injection port.

The order of appearance of the components in the effluent gases will correspond to the order of their boiling points—that is, *tert*-butylbenzene will be eluted first. Total elution time is about 3 minutes.

Identify each peak on the chart paper and indicate the attenuator setting at which each signal was obtained. Determine the areas of the *tert*-butylbenzene and 4-bromo-*tert*-butylbenzene peaks (Exp. 7, page 101) and calculate from the total areas the percentage of each component in the mixture of undistilled product. Record the percentages of each component in the mixture on your report form and attach the chromatogram.

Note 1
With the Gow-Mac 69-150 instrument (or its predecessors) the DC-200 column that comes with the instrument may be used for this experiment. Otherwise, columns that have been used in this experiment include a $\frac{1}{4}$-inch × 4-foot column packed with 10% DC-200 on Chromosorb P in the Gow-Mac instrument and a $\frac{1}{8}$-inch × 6-foot column packed with 8% SF96 on Chromosorb WHP (80/100) in the Carle instrument. Most columns packed with DC-200 (Dow), SF-96 (General Electric), SP-2100 (Supelco), or OV-101 (Foxboro/Analabs) should be satisfactory.

Preparation of 4-Bromo-*tert*-Butylbenzene

Reaction
equation

$C(CH_3)_3$ + Br_2 $\xrightarrow{Fe}$ $C(CH_3)_3$ ... Br

tert-Butylbenzene   4-Bromo-*tert*-butylbenzene

	tert-Butylbenzene	Bromine	Product
Quantities	31 mL (26.8 g)	8.2 mL (24 g)	_____
Mol. Wt.	_____	_____	_____
Moles	_____	_____	_____
Equivalents	_____	_____	_____

Sulfanilamide

The preparation of sulfanilamide illustrates the synthesis of an important medicinal product. The reaction sequence that leads to the preparation of this "sulfa" drug illustrates the sulfonation of an aromatic ring bearing a "protected" amino group.

Acetanilide (I) may be readily chlorosulfonated with chlorosulfonic acid to produce *p*-acetamidobenzenesulfonyl chloride (II). The sulfonyl chloride, when treated with ammonia, yields the sulfonamide, *p*-acetamidobenzenesulfonamide (III). When (III) is treated with aqueous hydrochloric acid only the carboxylic acid amide is hydrolyzed to produce *p*-aminobenzenesulfonamide (IV). *p*-Aminobenzenesulfonamide usually is called sulfanilamide because the common name of the parent acid is sulfanilic acid.

Precautions. *See HAZARD CATEGORY 2, page 4, regarding the safe handling of hydrochloric acid and ammonium hydroxide. See the **Caution** note on the following page regarding the properties and safe handling of chlorosulfonic acid.*

PROCEDURE The preparation of sulfanilamide is accomplished in three procedural steps. Complete Parts A and B of the sequence in the same laboratory period. You may defer Part C until the next laboratory period.

Caution! Perform Parts A and B of this experiment in a good exhaust hood. Chlorosulfonic acid is a highly toxic, corrosive chemical that you should handle with care. Use only dry graduates and flasks with this reagent, as it can react vigorously with water, to form sulfuric and hydrochloric acids. Should you spill chlorosulfonic acid on yourself, wash it off immediately with large amounts of water. Wear your safety glasses and disposable plastic gloves.

A. Preparation of *p*-Acetamidobenzenesulfonyl Chloride

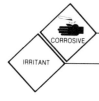

In a dry 50-mL Erlenmeyer flask place 12.5 mL (22.5 g, 0.21 mole) of chlorosulfonic acid. Cool the acid to 10–15° in a bath of ice water. *Clamp the flask in place.* Add 5 g of finely powdered, dry acetanilide in small portions (about 0.5 g each), using a stirring rod to mix the ingredients and keeping the temperature in the range of 10–20°. After all or most of the acetanilide has dissolved, remove the flask from the ice bath and clamp it in place on a water bath. Heat the solution gently on the water bath for 20 minutes to complete the reaction. Pour the reaction mixture slowly and carefully (avoid spattering) into a mixture of 100 g of ice and 50 mL of water. Rinse the flask with cold water and add the water to the main reaction mixture. Stir the mixture well, breaking up any lumps with a stirring rod. Filter the product on the Büchner funnel with suction and wash the product with 50 mL of cold water. Use the product immediately in the next step.

B. Preparation of *p*-Acetamidobenzenesulfonamide

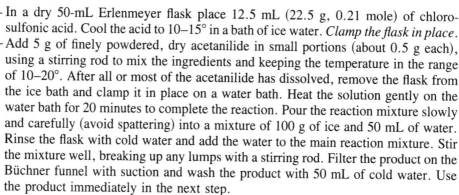

Transfer the crude, damp product from Procedure A to a 125-mL Erlenmeyer flask and add 20 mL of concentrated ammonium hydroxide, mixing well with a stirring rod. A reaction usually begins immediately and the mixture becomes warm. Incorporate the flask in a reflux setup and heat the mixture on a water bath for 15 minutes. Remove the flask and place it in an ice bath. When the mixture is cold, add 40–50 mL of dilute (6*N*) hydrochloric acid until the mixture is acid to litmus paper. Continue cooling in the ice bath until the mixture is thoroughly cold, and filter the product on the Büchner funnel with suction. Wash the product with 50 mL of cold water. The product may be used directly in the next step or allowed to dry in air.

C. Preparation of *p*-Aminobenzenesulfonamide

Transfer the crude product from Procedure B to a 125-mL flask and add a mixture of 10 mL of water and 5 mL of concentrated hydrochloric acid. Incorporate the flask in a reflux assembly and heat (gently at first to prevent charring) the mixture to boiling using a low flame under a wire gauze, a heating mantle, or an oil bath. Boil the solution for 20 minutes, during which time the solid should dissolve. Dilute the solution with an equal volume of water and, if discolored, add a small amount (matchhead size) of vegetable charcoal. Heat the solution to boiling and filter through a fluted filter into a 400-mL beaker. Stir in solid sodium carbonate in small portions until the solution is just alkaline to litmus. Chill the mixture thoroughly in the ice bath and collect the precipitate by suction filtration on the Büchner funnel. Wash the crystals with 25 mL of cold water and dry as much as possible on the suction filter. Recrystallize from water using about 10–12 mL of water per gram of crude product. Dry your product in the air and determine its melting point. Yield, approximately 2 g. Submit your product with your report.

REPORT **36**

Name ______ Section ______ Date ______

Sulfanilamide

Reaction equation

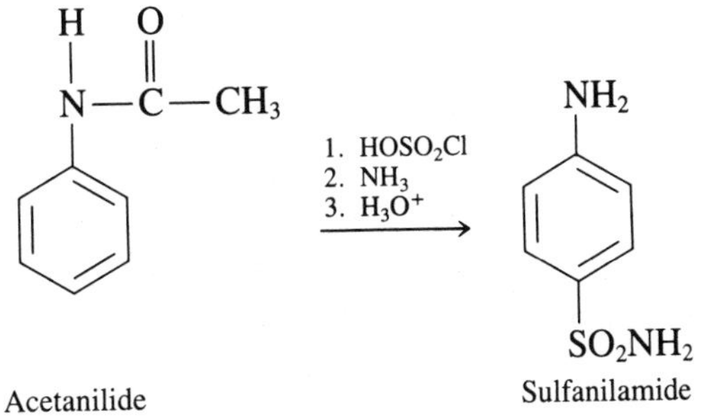

	Acetanilide	Sulfanilamide
Quantities	5.0 g	______ g
Mol. Wt.	______	______
Moles	______	______
Equivalents	______	______

Theoretical yield ______ g

Actual yield ______ g

Percentage yield ______ (based on acetanilide)

mp ______ °C

1. Write equations for the preparation of (a) *p*-toluenesulfonyl chloride from toluene and (b) N-methylbenzenesulfonamide from benzenesulfonyl chloride.

2. Based on your equation from Exercise 1, why was it necessary to protect the amino group in your preparation of sulfanilamide? What sort of product might result if the amino group were not protected?

3. In the hydrolysis step in the sulfanilamide preparation procedure, why is the carboxamide hydrolyzed rather than the sulfonamide?

4. The nmr spectrum of sulfanilamide appears below. Identify the protons responsible for each of the signals shown.

WAVELENGTH (MICRONS)

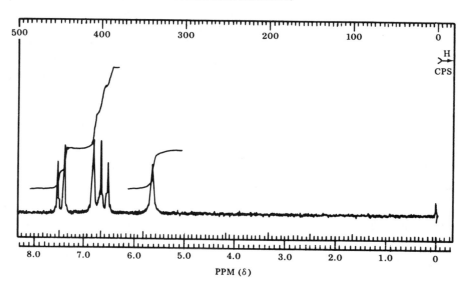

PPM (δ)

The Preparation of Pinacol Hydrate

The following experiments are related reaction sequences inasmuch as each begins with the bimolecular reduction of acetone to pinacol hexahydrate. Sequence I requires the preliminary drying of pinacol hydrate to the anhydrous form before it can be dehydrated by mild acid treatment to 2,3-dimethyl-1,3-butadiene. With the diene on hand, the stage is set for carrying out the Diels-Alder reaction by combining it with maleic anhydride to produce *cis*-4,5-dimethyl-1,2,3,6-tetrahydrophthalic anhydride. Sequence II begins with a change in the acid treatment above in order to favor a molecular rearrangement of pinacol to pinacolone over diene formation. A methyl ketone such as is exemplified by pinacolone suggests a route to pivalic acid via the haloform reaction. The two sequences may be shown in reaction form as follows:

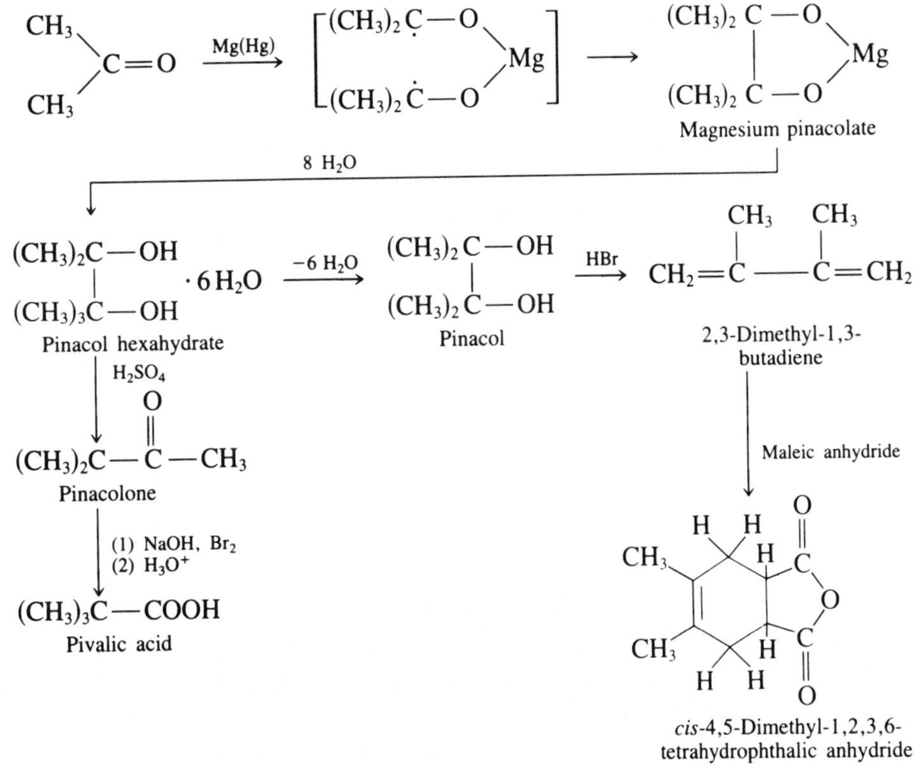

Pinacol hexahydrate

Pinacol

2,3-Dimethyl-1,3-butadiene

Magnesium pinacolate

Pinacolone

Pivalic acid

Maleic anhydride

cis-4,5-Dimethyl-1,2,3,6-tetrahydrophthalic anhydride

Each multistep sequence requires the practice of careful laboratory techniques and attention to detail but affords an opportunity for especially interested students and those of sufficient attainment to repeat some of the important accomplishments of earlier organic chemists.

The reactions show five different chemical changes and illustrate (1) a bimolecular reduction of a ketone to a diol, (2) the dehydration of a diol to a diene, (3) the Diels-Alder [4 + 2] cycloaddition, (4) the formation of a carbocation with a resulting molecular rearrangement, and (5) an oxidative degradation via the haloform reaction. Much interesting chemistry is involved here, but the following experiments will serve better to illustrate organic reactions and to provide instructional opportunity than to demonstrate any facility with which one organic compound may be converted to another. Unfortunately, the yields resulting in each of the above reactions are low, but by carrying out the sequences as a team involving two or more students, you may combine yields and obtain good positive results from each procedure.

Precautions. *This experiment requires the use of flammable liquid alkanes, acetone, tetrahydrofuran, and the very toxic divalent mercury compound mercuric chloride. See HAZARD CATEGORIES 1 and 3, page 4. In addition to handling the above chemicals safely, be reminded also that good laboratory practice requires the proper disposal of all potentially hazardous byproducts of a reaction. Most chemistry laboratories have made arrangements for the disposal of wastes containing mercury or mercury salts. Students must follow the directions their instructor provides for such disposal. In the event that facilities are not available for the proper disposal of mercury wastes, skip Experiment 37 and start the reaction sequence with Experiment 38 using commercially available pinacol or pinacol hexahydrate.*

PROCEDURE

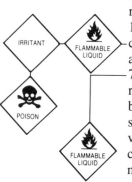

Assemble on a single ring-stand a *dry* 500-mL round-bottomed flask fitted with a reflux condenser bearing a drying tube and a funnel (Fig. 21.1). Place in the flask 100 mL of anhydrous tetrahydrofuran (Note 1) and dissolve in it 10 g of mercuric chloride. After the mercuric chloride has dissolved, add to the flask 12 g (0.5 gram atom) of bright magnesium turnings. Swirl the flask. Next add by way of the funnel 75 mL of anhydrous acetone in 5–10 mL increments. (Note 2) If the glassware and reagents used are dry, reaction is immediate and lively and will need to be moderated by periodic cooling of the flask in an ice-water bath. The reaction will proceed spontaneously for the next 30–45 minutes and should be allowed to continue as vigorously as possible without any undue amount of cooling. Allow refluxing to continue until it subsides of its own accord; then, using a steam bath or a heating mantle, reflux for an additional hour. The reaction mixture becomes progressively more viscous and turns into a gelatinous mass by the end of the reflux period. Stirring is difficult, but attempt from time to time to break up the mass by vigorously swirling the entire assembly. Hydrolyze the magnesium pinacolate by adding 150 mL of saturated sodium chloride solution (brine), and heat and swirl for 30 minutes. Allow the mixture to settle, then carefully decant the supernatant solution into an Erlenmeyer flask and set aside. Add 50 mL of ordinary tetrahydrofuran and 50 mL of brine to the grey precipitate that remains in the flask and heat for 15 minutes. Again, decant this extract and add it to the main solution. Filter the residue on a Büchner funnel and add the filtrate to the combined solutions obtained by decantation. (Note 3) Place the residue in a container especially reserved for toxic wastes. Separate the organic layer from the brine. Discard the brine and reduce the volume of the combined solutions by removing approximately 100 mL of solvent by distillation. Precipitate the pinacol hydrate from the residual solution by adding 150 mL of petroleum ether or hexane and cooling in an ice-water bath. (**Caution!** *Flammable solvent.*) Collect the product on a Büchner funnel and press the crystals as dry as possible between layers of filter paper. The product is sufficiently pure to be used in subsequent preparations. Store in a tightly stoppered container. (Note 4) Yield, 16 g; mp 46–47°.

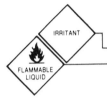

Note 1 When exposed to air repeatedly as in the case of old, nearly empty bottles of the reagent, tetrahydrofuran readily forms peroxides and may pose an explosive hazard if old tetrahydrofuran is used as starting material. A fresh bottle of tetrahydrofuran that has been stabilized should be used for this experiment. The presence of peroxides may be tested as described in Note 1, p. 57.

Note 2 The glassware and reagents used in this preparation must be dry, as in the preparation of a Grignard reagent.

Note 3 Filtration is very slow but may be expedited by the use of Celite or Filter-aid.

Note 4 Pinacol hydrate sublimes readily.

Name   Section   Date

The Preparation of Pinacol Hexahydrate

Reaction
equation

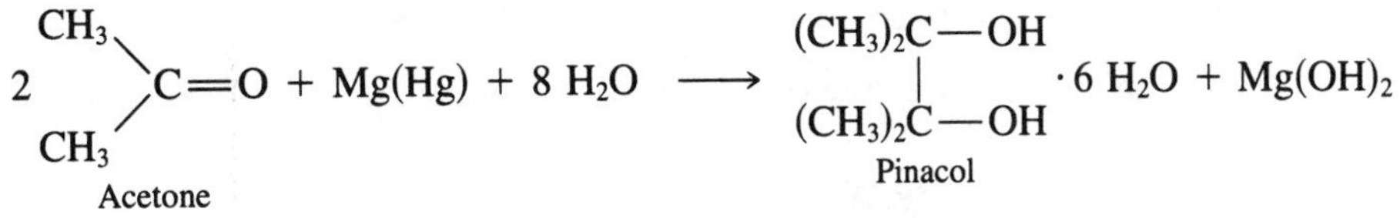

$$2 \underset{CH_3}{\overset{CH_3}{\diagdown}}C{=}O + Mg(Hg) + 8\ H_2O \longrightarrow \begin{matrix} (CH_3)_2C{-}OH \\ | \\ (CH_3)_2C{-}OH \end{matrix} \cdot 6\ H_2O + Mg(OH)_2$$

Acetone   Pinacol

	Acetone	Magnesium	Pinacol hydrate
Quantities	_____	_____	_____
Mol. Wt.	_____	_____	_____
Moles	_____	_____	_____
Equivalents	_____	_____	_____

Theoretical yield _____ g (based on gram-atoms of Mg taken)

Actual yield   _____ g

Percentage yield _____

1. Aldehydes and ketones may be reduced in a variety of ways. Write equations to show the products that result when the following reducing reagents are used on acetone: (a) Zn(Hg), HCl; (b) $NaBH_4$; (c) H_2, Pd; (d) Na.

2. Draw the structure of, and name the product that results from, the bimolecular reduction of acetophenone.

3. When the compound of Question 2 is treated with sulfuric acid, two products are possible, but one predominates. Which is the major product and why is its formation favored?

4. Benzophenone is reduced to benzopinacol when a solution in 2-propanol is exposed to sunlight. Show how this reduction takes place by way of a free-radical mechanism.

5. What different products could result if 1-phenyl-1,2-propanediol were to undergo a pinacolic rearrangement? What factors determine which product is the likely one?

*6. How many peaks would the nmr spectrum of pinacolone show? What is the multiplicity of the peaks—i.e., are the peaks singlets, doublets, etc? What is the ratio of protons in the peaks? Relative to TMS, at what δ values do the peaks appear? (*Note:* Consult a TABLE of proton nmr shift values.) Draw an nmr spectrum for pinacolone that embodies the answers to each of the previous questions.

The Diels-Alder Reaction: The Preparation of *cis*-4,5-Dimethyl-1,2,3,6-tetrahydrophthalic Anhydride

One of the most important reactions of conjugated dienes is the **Diels-Alder reaction**—a [4 + 2]-cycloaddition that these unsaturated compounds undergo when they react with active ethylenic or acetylenic compounds.

Diene

Dienophile   Adduct

The reaction involves a conjugated *diene* and a *dienophile* (Greek, diene loving) to yield a cyclohexene derivative. No catalyst is required, but the reaction is facilitated when the dienophile has an electron-withdrawing group conjugated with a multiple bond such as is found in an α,β-unsaturated carbonyl or nitrile. The reaction is stereospecific in that the configurations of both the diene and the dienophile are retained in the adduct.

In the present experiment we will illustrate the Diels-Alder reaction by combining 2,3-dimethyl-1,3-butadiene with maleic anhydride to give the cyclic adduct, *cis*-4,5-dimethyl-1,2,3,6-tetrahydrophthalic anhydride.

The Preparation of Anhydrous Pinacol

The water of hydration must be removed first before pinacol hydrate can be converted into 2,3-dimethyl-1,3-butadiene. The dehydration step is accomplished by refluxing the hydrate with an organic solvent that forms an azeotrope with water. Although pinacol hydrate is but sparingly soluble in cyclohexane, water is even less so. By heating the hydrate with cyclohexane under reflux and utilizing a water trap in our assembly as illustrated in Figure 38.1 we can effectively dehydrate the pinacol. As the azeotrope reaches the condenser, the solubility of water in cold cyclohexane is but a very small fraction of its solubility in hot cyclohexane, and water separates as a second phase, dropping to the bottom part of the water trap. Cyclohexane fills the upper portion of the trap and overflows back into the distilling flask. A calibration of the water trap permits a close approximation of the completeness of dehydration if the amount of material to be stripped of water is first weighed. However, the calibration feature, while convenient, is not necessary because it usually is quite obvious when droplets of water no longer form in the trap.

Precautions. *This experiment requires the use of cyclohexane, tetrahydrofuran, hydrobromic acid, and maleic anhydride. See HAZARD CATEGORIES 1, 2, and 8, page 4.*

Figure 38.1   Water separator assemblies. (a) Improvised trap from standard taper ware; (b) Dean-Stark trap.

Indicates placement of clamp

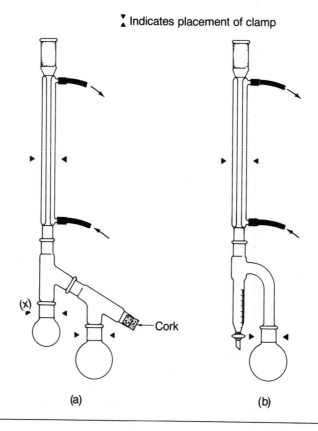

(a)   (b)

PROCEDURE

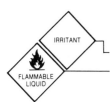

Assemble a reflux apparatus with a Dean-Stark water trap interposed between a 250-mL distilling flask and the condenser, or, if this specially designed trap is not available, improvise a water trap by utilizing connecting adapters and a 25-mL flask as shown in Figure 38.1(a). (Note 1)

Weigh and place the entire yield of pinacol hydrate obtained from Experiment 37 into the distilling flask, add 100 mL of cyclohexane, and heat to reflux using a heating mantle or a steam bath. (*No flames!*) Reflux until approximately 0.5 mL of water is collected for each gram of pinacol hydrate used. When no more water appears to pass over, discontinue refluxing, disconnect the water trap, and discard the water and cyclohexane collected. Rearrange your apparatus for distillation and distill again using a heating mantle or a steam bath. After removing the cyclohexane (bp 81°), change receivers and, *if others near you no longer are distilling cyclohexane*, use a Bunsen burner adjusted to a low flame, a heating mantle, or an oil bath to collect all product boiling between 170–175° as anhydrous pinacol. Anhydrous pinacol will solidify upon cooling (mp 38°).

Note 1   To avoid any leakage at joint (x) in Figure 38.1a, close the clamp supporting the condenser only tightly enough to keep the condenser in place yet allow its weight to bear upon the connecting adapter below and indirectly upon the water-receiving flask. In addition, a rubber band looped around the adapter above the side-arm and around the clamp supporting the water receiver will help prevent leakage. All joints should be *lightly* greased.

PART I   The Preparation
of 2,3-Dimethyl-1,3-butadiene

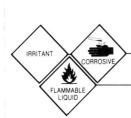

Assemble an apparatus for a small-scale fractional distillation such as that illustrated in Figure 38.2, using a 50-mL round-bottomed flask as the boiling flask and a 25-mL flask as the receiver. Improvise a short, bead-packed fractionating column from a drying tube or from a modified Claisen adapter. Place 11.8 g (0.10 mole) of anhydrous pinacol and 1.0 mL of 48% hydrobromic acid in the flask. Add a Boileezer and distill, using a steam bath, oil bath, or heating mantle. Insulate the packed column with aluminum foil to prevent too much heat loss. Collect all distillate that passes over below 90°. The diene has a boiling point of 76°, and pinacolone, some of which also forms, has a boiling point of 106°. Continue distillation until the temperature cannot easily be kept below 90° even though a small volume (approximately 1–2 mL) remains in the distillation flask. Place the distillate in a small separatory funnel and wash with 10 mL of cold water. Separate the organic layer and dry it over a few granules of anhydrous calcium chloride while you prepare for Part II. Transfer the residual liquid that remains in the distillation flask to a 25-mL Erlenmeyer, stopper, and label it "Crude pinacolone." Save for a subsequent experiment. *Use the diene during the laboratory period in which it is prepared.*

Figure 38.2 (a) Small-scale fractional distillation assembly showing Claisen adapter as a packed column; (b) glass-jointed drying tube as a packed column.

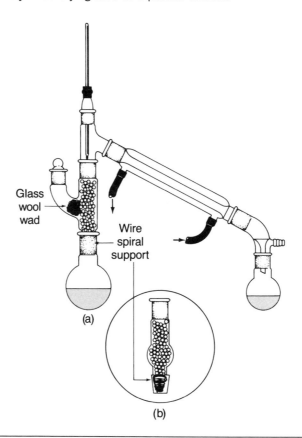

Glass wool wad

Wire spiral support

(a)

(b)

PART II   Preparation of *cis*-4,5-Dimethyl-1,2,3,6,-tetrahydrophthalic Anhydride

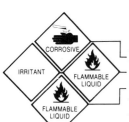

In a 50-mL round-bottomed flask dissolve 5 g (0.05 mole) of finely pulverized maleic anhydride in 15 mL of anhydrous tetrahydrofuran. Warm if necessary to effect solution. Cool, then add 5 mL (4.2 g., 0.05 mole) of 2,3-dimethyl-1,3-butadiene. Swirl to mix. If the entire yield of Part I is taken, the diene should first be weighed and 1.17 g of maleic anhydride dissolved for each gram of diene taken. Shake the flask to mix thoroughly and note that the mixture becomes quite warm. Attach a reflux condenser to the flask and reflux for 30 minutes. Cool the reaction mixture in an ice-water bath, agitate, and scratch the sides of the flask with a stirring rod to initiate crystallization. The formation of crystals is sometimes a little slow. Collect your product on a Hirsch funnel, wash with 10 mL of cold water, and dry. Yield, 3 g; mp 78°.

Name ________ *Section* ________ *Date* ________

Preparation of *cis*-4,5-Dimethyl-1,2,3,6-tetrahydrophthalic Anhydride

Reaction equations

PART I

$$\underset{\text{Pinacol}}{(CH_3)_2C(OH){-}C(OH)(CH_3)_2} \xrightarrow{48\%\ HBr} \underset{\text{2,3-Dimethyl-1,3-butadiene}}{CH_2{=}C(CH_3){-}C(CH_3){=}CH_2} + 2\,H_2O$$

	Pinacol	Diene
Quantities	11.8 g	________
Mol. Wt.	________	________
Moles	________	________
Equivalents	________	________

Theoretical yield ________ g

Actual yield ________ g

Percentage yield ________

bp ________°C

PART II

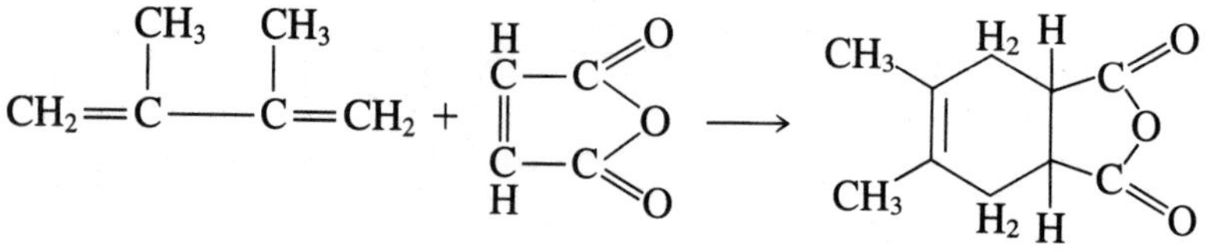

Maleic
anhydride

cis-4,5-Dimethyl-
1,2,3,6-tetrahydrophthalic
anhydride

	Diene	Anhydride	Adduct
Quantities	_____	_____	_____
Mol. Wt.	_____	_____	_____
Moles	_____	_____	_____
Equivalents	_____	_____	_____

Theoretical yield _____ g

Actual yield _____ g

Percentage yield _____

mp _____ °C

1. How could you prepare benzoic acid using a Diels-Alder reaction as the first step in a three-part sequence?

2. Write structures for the adducts that result when the following compounds react with 2,3-dimethyl-1,3-butadiene.
   (a) acrylonitrile
   (b) acrolein
   (c) crotonaldehyde
   (d) *p*-benzoquinone

3. It was suggested in the procedure that 2,3-dimethyl-1,3-butadiene be used within the same laboratory period during which it is made. Why?

The Preparation of 2,2-Dimethylpropanoic Acid (Pivalic Acid)

One of the more interesting reactions exhibited by vicinal glycols is the facility with which they undergo intramolecular acid-catalyzed rearrangements. The mechanism of the rearrangement is fairly well established as one initiated by a protonation of one of the hydroxyl groups, followed by a dehydration and formation of a carbocation. Migration of a neighboring group *with its pair of electrons* to the electron-deficient carbon then results in a rearrangement of the carbon skeleton.

The reaction is not limited to compounds that bear only hydroxyl groups on adjacent carbons but also occurs when amino and alkoxy groups appear on neighboring carbon atoms. All rearrangements of this type are referred to as pinacolic rearrangements. In addition to alkyl groups, migrating groups include aryl and hydrogen. From a number of studies of the reaction, the migratory aptitude of groups follows the order $Ar:^- > R:^- > H:^-$.

In the present experiment we will employ the pinacol–pinacolone rearrangement to prepare our starting material for synthesizing pivalic acid.

PART I   The Preparation of Pinacolone

> **Precautions.** *This experiment requires the use of ether, bromine, sulfuric acid, and sodium hydroxide. See HAZARD CATEGORIES 1, 2, and 3, page 4.*

PROCEDURE

In a 100-mL round-bottomed flask equipped with an adapter and condenser set for distillation place 65 mL of 6N sulfuric acid and 11.3 g (0.05 mole) of previously prepared pinacol hydrate. Heat the mixture with a Bunsen burner and collect the distillate in a graduate cylinder. Continue the distillation until the upper organic layer in the graduate cylinder reaches a maximum value and only one phase remains in the distilling flask. The collection of 25 mL of distillate should be adequate. Transfer the distillate to a separatory funnel, separate, and discard the lower aqueous layer. Store the pinacolone in a small Erlenmeyer flask over a few granules of anhydrous calcium chloride, stopper, and save for Part II. (Note 1)

PART II   The Preparation of Pivalic Acid

In a 125-mL Erlenmeyer flask dissolve 4 g of sodium hydroxide in 35 mL of water and cool the solution in an ice-water bath. To the cold sodium hydroxide solution add 9.6 g (3.2 mL, 0.06 mole) of bromine from the bromine buret located in the hood. Swirl the flask until the bromine dissolves and the color of the solution is a light yellow. After the bromine has dissolved add 2.0 g (2.2 mL, 0.02 mole) of pinacolone. Stopper the flask and shake vigorously and intermittently for 30 minutes. Release the stopper slightly to vent any pressure after each shaking. The flask will become warm and the yellow color of the reaction mixture will fade completely. Transfer the mixture to a 100-mL round-bottomed flask and, using a Bunsen burner, remove the bromoform by steam distillation. Collect about 15 mL of distillate. Cool the distillation flask by immersing it in ice water, then transfer the cold contents to a separatory funnel.

Extract the aqueous solution once with 15 mL of ether. Discard the ether layer and collect the aqueous layer in a small Erlenmeyer flask. Set the flask in a pan of hot water and swirl periodically to drive out any dissolved ether. When you can no longer detect the odor of ether, set the flask in an ice bath and acidify the solution to Congo Red paper (Note 2) by the dropwise addition of concentrated sulfuric acid. If the solution turns yellow upon addition of acid, decolorize with a few drops of saturated sodium bisulfite solution. Continue to cool until crystallization is complete. Pivalic acid crystals will appear as prismatic needles. Collect on a Hirsch funnel. Yield, 1–1.2 g; mp 34°.

Note 1 A small amount of 2,3-dimethyl-1,3-butadiene (bp 76°) will also appear in the distillate with pinacolone and may be removed by fractional distillation on a micro-scale. A product distilling between 103–110° should be collected as pinacolone (bp 106°). However, the presence of the diene in pinacolone does not interfere with the haloform reaction that yields pivalic acid because any neutral organic compounds formed by bromination are removed in the extraction step before the acid is precipitated.

Note 2 The use of Congo Red paper and substitutes for it are discussed in Note 5 of Experiment 5, page 77.

Name   Section   Date

The Preparation of 2,2-Dimethylpropanoic Acid (Pivalic Acid)

Reaction equations

$$(CH_3)_2C{-}OH \\ \quad\quad\quad\quad | \quad\quad \cdot 6\ H_2O \xrightarrow{H_3O^+} (CH_3)_3C{-}\overset{\overset{\textstyle O}{\|}}{C}{-}CH_3 + 7\ H_2O \\ (CH_3)_2C{-}OH$$

Pinacol hexahydrate   Pinacolone

$$(CH_3)_3C{-}\overset{\overset{\textstyle O}{\|}}{C}{-}CH_3 + 3\ Br_2 + 4\ NaOH \longrightarrow$$

$$(CH_3)_3C{-}COO^-\ Na^+ + 3\ NaBr + 3\ H_2O + CHBr_3$$

$$(CH_3)_3C{-}COO^-\ Na^+ + H_2SO_4 \longrightarrow (CH_3)_3C{-}COOH + NaHSO_4$$

Pivalic acid

	Pinacolone	Pivalic acid
Quantities	2 g	_____
Mol. Wt.	_____	_____
Moles	_____	_____
Equivalents	_____	_____

Theoretical yield _____ g

Actual yield _____ g

Percentage yield _____

mp _____ °C

1. Devise a synthesis for pinacolone that involves acetone but not pinacol.

2. Pinacolone will give a positive haloform reaction to yield chloroform or bromoform but fails to give a positive iodoform reaction. Why?

3. What property does pivalic acid have in common with other aliphatic acids of four or more carbons?

4. Devise a synthesis of pivalic acid that would not require the use of the haloform reaction in any step.

5. If you attempted a nitrile synthesis followed by hydrolysis in your solution to Question 4, your yield of pivalic acid would be very small indeed. What would be your principal product? Why?

Appendices

Infrared and Nuclear Magnetic Resonance Spectra of Compounds Listed in Table 16.1

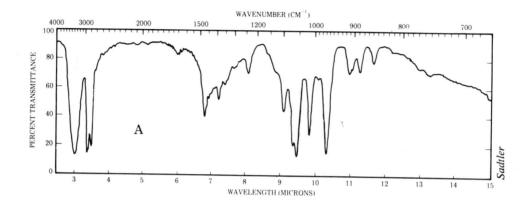

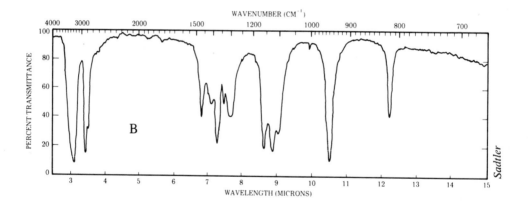

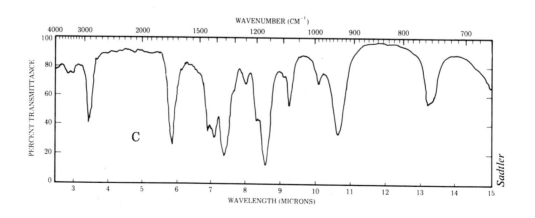

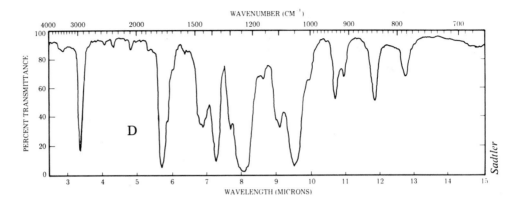

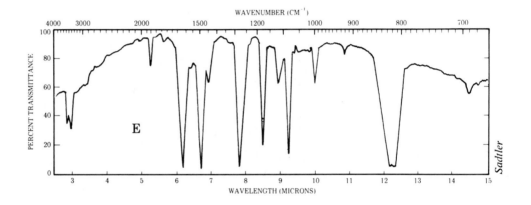

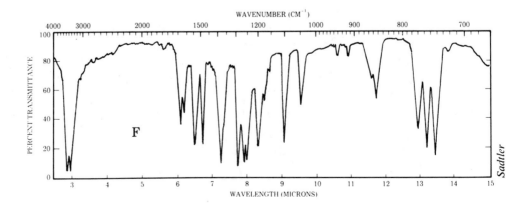

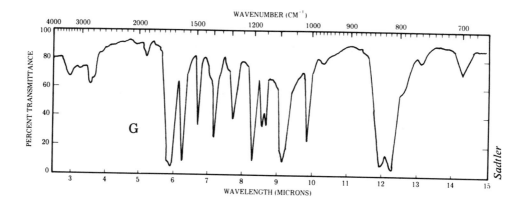

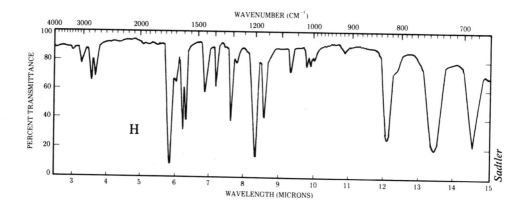

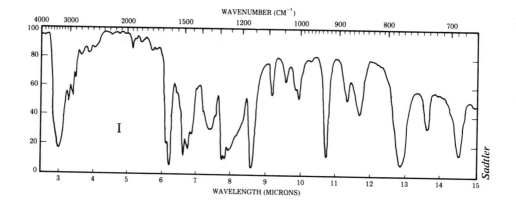

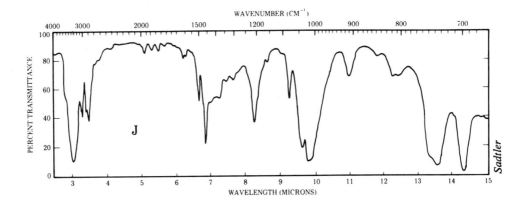

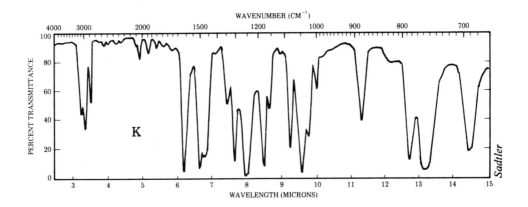

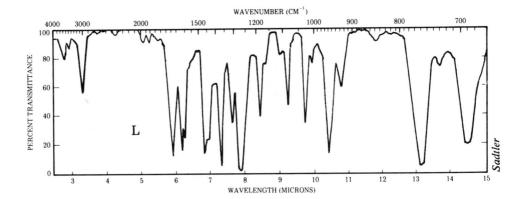

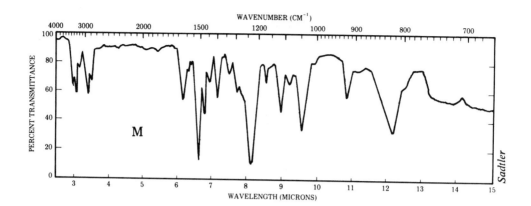

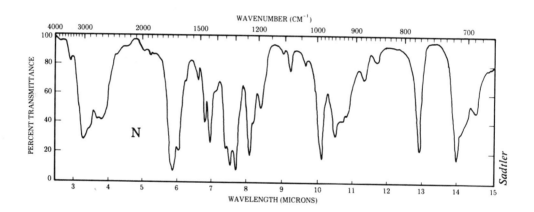

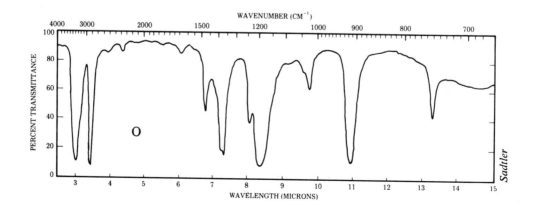

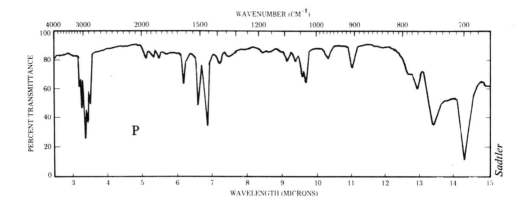

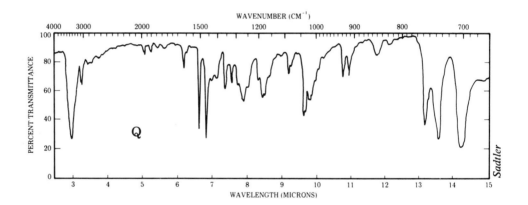

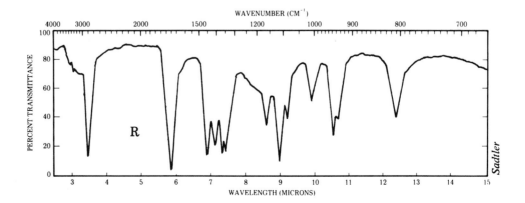

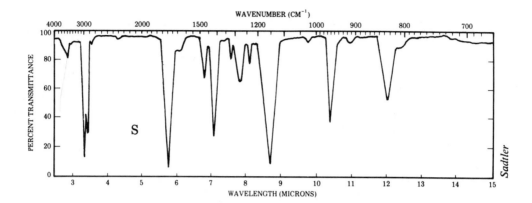

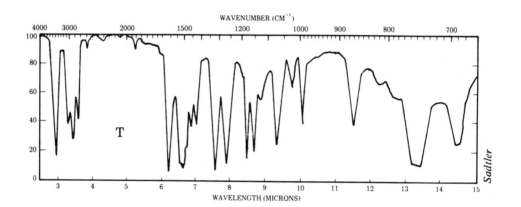

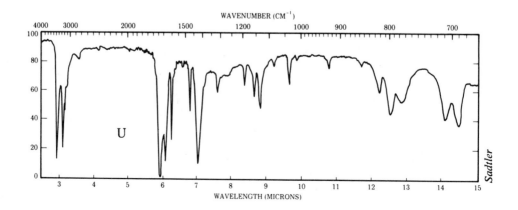

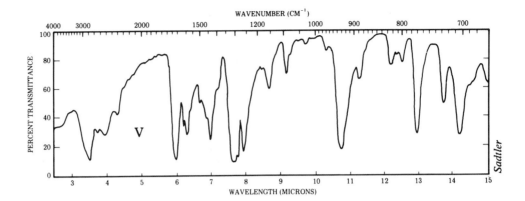

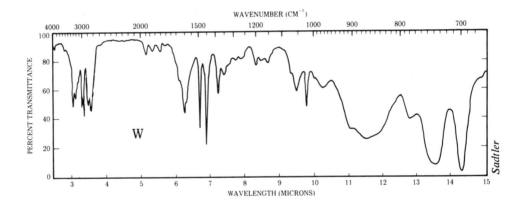

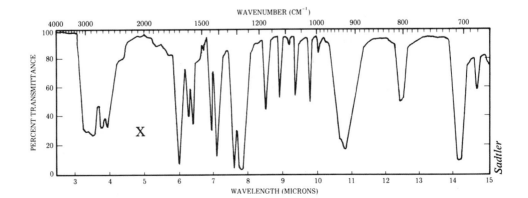

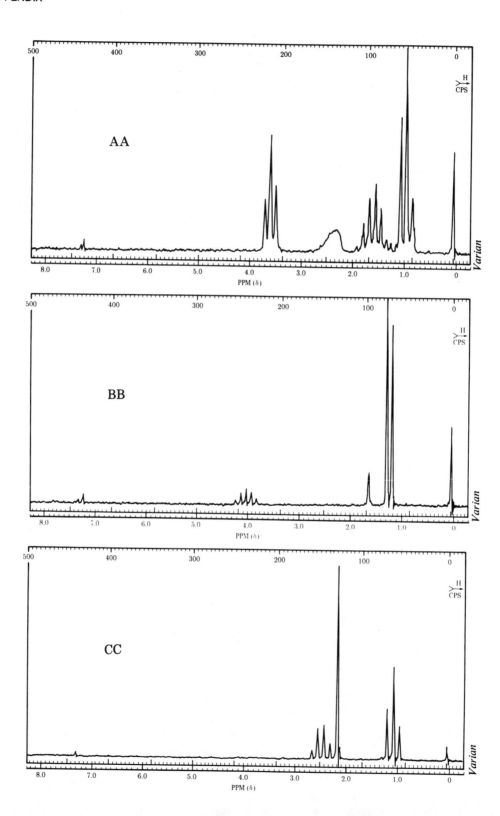

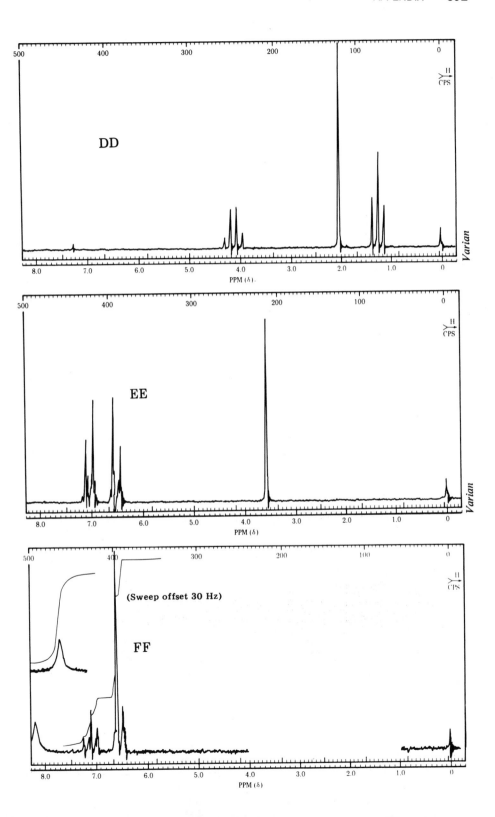

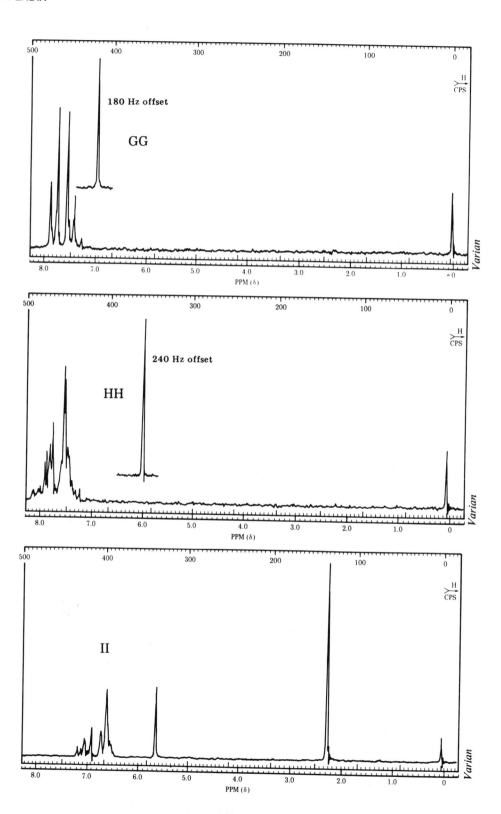

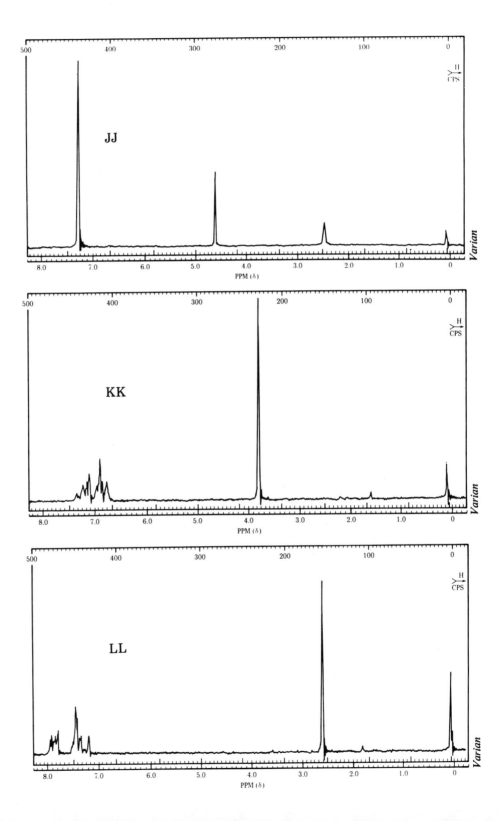

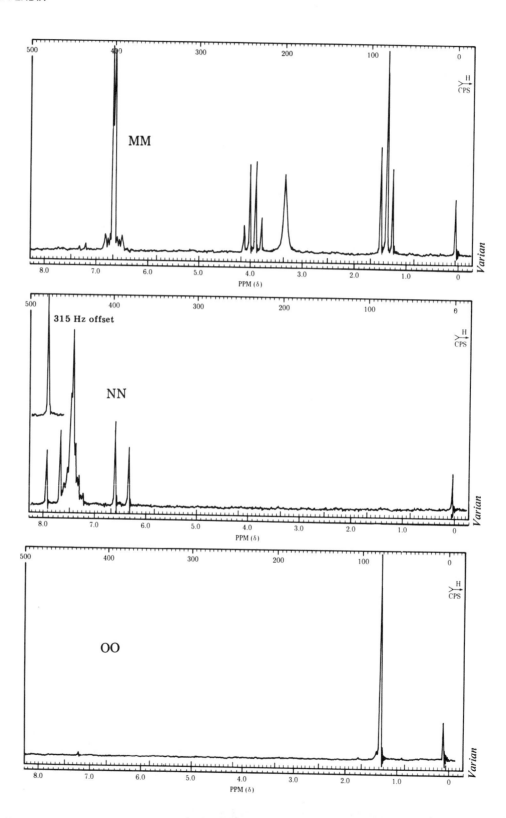

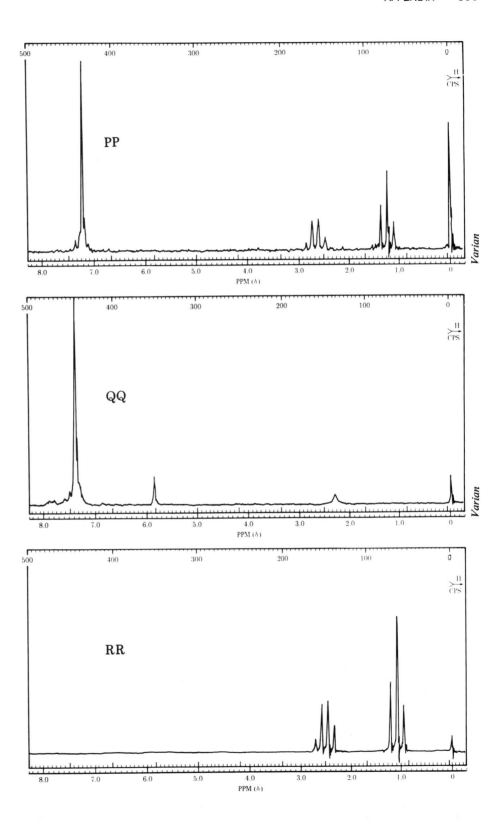

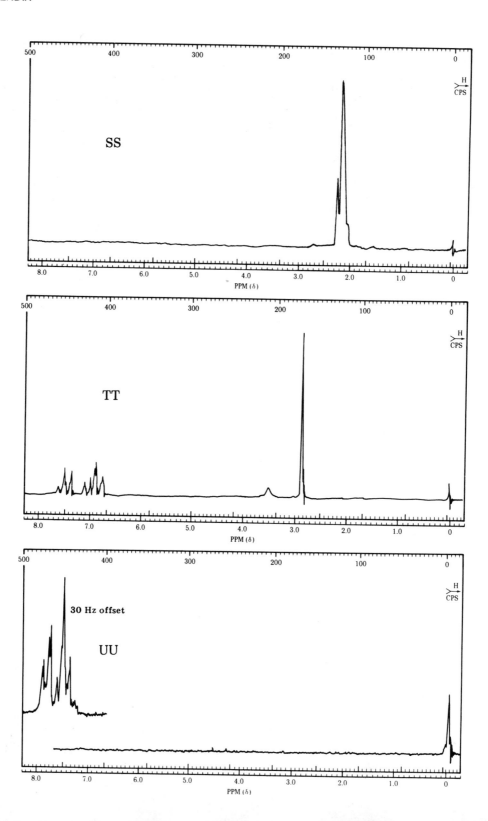

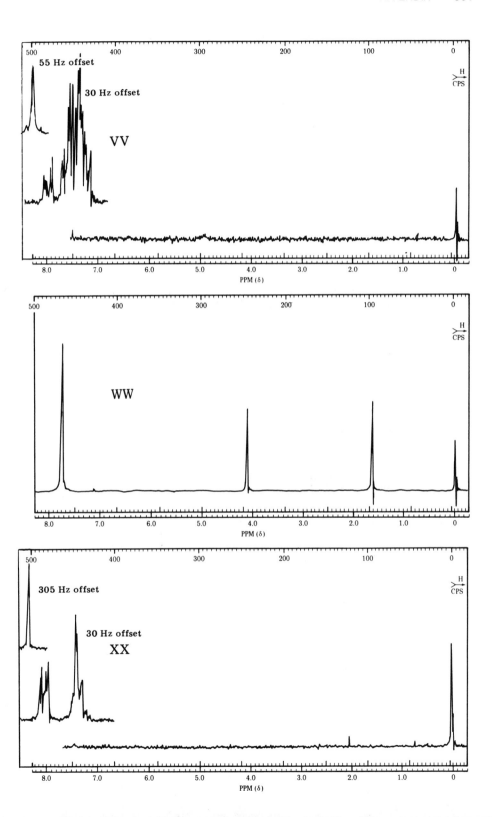

Key to the Identity of IR and NMR Spectra of Compounds Listed in Table 16.1.

IR	NMR	Compound
A	AA	1-Propanol
B	BB	2-Propanol
C	CC	2-Butanone
D	DD	Ethyl acetate
E	EE	4-Chloroaniline
F	FF	1, 2-Dihydroxybenzene (Catechol)
G	GG	4-Chlorobenzaldehyde
H	HH	Benzaldehyde
I	II	*m*-Cresol
J	JJ	Benzyl alcohol
K	KK	Anisole (methyl phenyl ether)
L	LL	Acetophenone (methyl phenyl ketone)
M	MM	4-Ethoxyaniline (*p*-phenetidine)
N	NN	*trans*-Cinnamic acid
O	OO	2-Methyl-2-propanol (*tert*-butyl alcohol)
P	PP	Ethylbenzene
Q	QQ	Diphenylmethanol (benzhydrol)
R	RR	3-Pentanone
S	SS	Cyclopentanone
T	TT	N-Methylaniline
U	UU	Benzamide
V	VV	*o*-Benzoylbenzoic acid
W	WW	Benzylamine
X	XX	Benzoic acid

Index